Creating a Lean and Green Business System
Techniques for Improving Profits and Sustainability

D0140771

Creating a Lean and Green Business System
Techniques for Improving Profits and Sustainability

Keivan Zokaei • Hunter Lovins
Andy Wood • Peter Hines

CRC Press
Taylor & Francis Group
Boca Raton   London   New York

CRC Press is an imprint of the
Taylor & Francis Group, an **informa** business

A PRODUCTIVITY PRESS BOOK

CRC Press
Taylor & Francis Group
6000 Broken Sound Parkway NW, Suite 300
Boca Raton, FL 33487-2742

© 2013 by A. Keivan Zokaei
CRC Press is an imprint of Taylor & Francis Group, an Informa business

No claim to original U.S. Government works

Printed on acid-free paper by CPI Group (UK) Ltd, Croydon
Version Date: 20130425

International Standard Book Number-13: 978-1-4665-7112-9 (Paperback)

Library of Congress Cataloging-in-Publication Data

Zokaei, A. Keivan.
    Creating a lean and green business system : techniques for improving profits and sustainability / Keivan Zokaei, Hunter Lovins, Andy Wook, Peter Hines.
      pages cm
      Includes bibliographical references and index.
      ISBN 978-1-4665-7112-9 (pbk. : alk. paper)
      1. Industrial management--Environmental aspects. 2. Organizational effectiveness. 3. Business logistics. I. Title.

    HD30.255.Z65 2013
    658.4'083--dc23   2013003039

Visit the Taylor & Francis Web site at
http://www.taylorandfrancis.com

and the CRC Press Web site at
http://www.crcpress.com

Contents

Foreword by Jonathon Porritt

All is not well with the world. The bad things (such as accelerating climate change, social injustice, resource shortages and so on) are getting worse faster. And the best efforts of those seeking to address those bad things are falling ever shorter. Any assumptions we might once have had as to how we get out of this mess are proving to be pretty much worthless.

Let's just test one of those assumptions: that it is elected governments (with their democratic mandates) that are best placed to broker a fairer balance between the interests of people today and the interests of future generations. Sadly, that powerful idea of justice between generations (which lies at the heart of the most familiar of all the definitions of sustainable development: development that meets the needs of the present without compromising the ability of future generations to meet their own needs) would appear to be less available to us today than it was back in 1992 at the time of the Earth Summit in Rio de Janeiro.

We know that because of what happened at the Rio+20 Summit in June 2012. Here we witnessed 190 world leaders, in the same place at the same time, committing their governments to a communiqué that is exceptional only in its vacuous banality. However uncomfortable it may be, we have to be honest in recognizing the implications of the Rio+20 fiasco: almost all governments today are either incapable or unwilling to take the lead on establishing the basic conditions for a more sustainable world, and will always prefer their own short-term interests and those of their voters over the hard graft of negotiating a more sustainable economy.

Worse yet, because the United Nations (UN) is essentially a creature of those same dysfunctional governments, it too has proved to be incapable of taking such a lead.

In this contested territory of intergenerational justice, governments have essentially opted out of leading their electorates and have settled for following those who are still prepared to do more of the heavy lifting—from whichever sector they may come.

There was one symbolic manifestation of all this at the Rio+20 conference, concerning the Consumer Goods Forum (CGF)—a coalition of more than 40 global retailers and fast-moving consumer goods companies. The CGF was extremely active in the run up to Rio in getting its members to commit to sourcing strategies that would help to protect the world's remaining rainforests.

This was a voluntary initiative, to be sure, and therefore, like all such voluntary initiatives, unenforceable. But this one has real substance and was well received, both by NGOs (nongovernmental organizations) and by the business community at large. Governments were also supportive, given how little success they've had to date in protecting those rainforests.

Indeed, about the only vaguely useful thing the US government did during Rio+20 was to commit publicly to supporting those companies in whatever initiatives they now take to honor the CGF's voluntary commitment! Why do something yourself (in government) when you can get others (in business) to do it for you?

I believe this "governance shift" (with governments stepping back and business stepping forward) is both crucially important and deeply problematic. It's hard to say whether it's a "good thing" or a "bad thing" in terms of getting our economies onto a more sustainable path. But I have to admit that my intense apprehension at the prospect of profit-maximizing multinationals taking on more and more responsibility, without so much as a sliver of democratic accountability, is outweighed by my increasingly urgent desire to see anybody step up to the plate.

These governance issues form an important part of the hinterland behind *Creating a Lean and Green Business System*, which looks closely at the role of progressive companies in rising to today's sustainability challenges. It provides a sharp, evidence-based celebration of some of the best companies out there today, who are contributing more than their fair share to the pursuit of a more sustainable world.

Much of this is often captured under the notion of "the business case for sustainable development," first elaborated by the World Business Council for Sustainable Development in the run up to the Rio Earth Summit twenty years ago.

The true significance of that business case (i.e., things that are good for business as well as good for people and the planet) is highlighted throughout. Take the case of DuPont, which committed itself to a 65% reduction in greenhouse gas emissions over a 10-year period up to 2010, netting the company billions of dollars every year through energy efficiency savings. In 2007, that savings was $2.2 billion—a year in which its total declared profits were not much more than $2 billion!

No surprise then that those companies that are recognized leaders in ESG (environment, society, and governance) issues have regularly outperformed the MSCI (Metals Service Center Institute) by more than 20% since 2005.

But the authors of this book remind us that there's so much more to the business case than simply cutting costs to protect the bottom line. Indeed, they suggest that those companies that obsess about waste and cost reduction, to the exclusion of all else, will *never* secure the profound cultural change inside their companies on which true corporate sustainability increasingly depends. For them, being lean and truly green is not just about the dollars saved, but about "continuous value enhancement for the whole of society."

All the case studies are carefully designed to promote that kind of shared learning. The Toyota case study is particularly compelling, not least because so many people would appear to have seized on the cost-cutting aspects of the Toyota Production System without properly understanding the cultural context in which those efficiency gains have been achieved.

That context is all important. For instance, while dependent on automation (as *all* car manufacturers have been for years), Toyota talks more about *autonomation*, ensuring that employees are not subjected to mindless repetitive tasks and are fully engaged in both the design and the monitoring of the Toyota Production System.

The authors were particularly struck by Toyota's use of the notion of *Monozukuri*—manufacturing in harmony with nature and adding value for society—or, as the authors put it, "the older sister of sustainable manufacturing." *Creating a Lean and Green Business System* is therefore as much about culture, values, and system change as it is about efficiency and the bottom line. And that kind of emphasis provides some important insight into the broader debate about the role of business in society today. The kind of values orientation highlighted in most of the case studies and examples is what builds trust between those companies and their customers—and that kind of trust is essential as business takes on a more and more significant role in mapping out the path to a more sustainable world.

Jonathon Porritt
Founder Director of Forum for the Future
http://www.forumforthefuture.org.

Foreword by Doc Hall

About 20 years ago I noticed that many of *IndustryWeek's* best plant applicants in the United States showed reductions in energy usage and material usage. Few mentioned any serious environmental program, but they were seriously into lean transformation. This made sense. Reductions came from eliminating unnecessary space, material handling, and rework, thus reducing materials scrapped and the energy to process it. Water usage was not measured. As the years progressed, these trends became more noticeable.

In the past ten years, a few companies began blending green into their lean. The most persistent ones had results similar to companies documented in this book. Unfortunately, we also noticed a silo effect, inside companies and in general. Lean programs and environmental programs were usually separate organizational silos that sometimes came into conflict with each other. Attempts to expand the impact by getting more lean leaders and environmental leaders to work together gained very little traction.

Creating a Lean and Green Business System updates similar initiatives and proposes how to move forward. It has a framework, an overview of tools, and case studies showing how companies can not only reduce environmental footprints, but also improve service to customers at the same time. This is a happy confluence. If costs go down and customer satisfaction goes up, good business results will follow.

Far too many general managers have not figured this out. It is counterintuitive to old business models forged from a financial view; surely paying to clean up a mess will cost more. Of course it does, if that is as far as the thinking goes. However, operating so as to prevent messes reduces cost; neither the company nor anyone else has to clean up something that didn't happen. Avoiding future messes both reduces present costs and minimizes future liabilities, many of which we cannot foresee.

"Accounting thinkers" cannot easily grapple with this. Transaction systems do not measure things that did not happen. Accounting can only project the value of preventing something from happening if somebody hangs a price tag and probability on consequences. (Business in general is not hopeless in this regard. Insurance companies attach a value to many foreseeable calamities.)

Business leaders fear this emerging world if they do not know what to do. Prior experience teaches them to identify only value at risk—events that they can foresee and quantify that could affect their profit and loss statements. Flocks of "black swans" out on the horizon defy conventional risk analysis. Hence, they tend to be ignored for practical purposes.

Stop being afraid and seize new opportunities. *Creating a Lean and Green Business System* is a good guide on how to start—and keep going. Companies that ignore both lean and green had better have a monopoly with a big margin to cover the waste financially. That in turn begs a further fundamental question—whether the entire business is merely waste that is financially viable at the moment. All businesses have limited lifetimes. If they ignore their own long-term health as well as ours, lifetimes are more likely shorter.

None of the example cases represent an ideal attainable state. All cases end with lessons learned and caveats that the organization can still do much more. An ideal of zero human impact on the planet is impossible. We must get much closer to zero, but all humans and other living creatures consume resources and exhale carbon dioxide just by being here. We are symbiotic with our environment, which is ever changing. All our present states are temporary. All our proposed future states are movable targets. New problems and challenges will emerge. All we can do is try to create a good quality of life using minimal natural resources, while keeping the world in a "Goldilocks Zone" in which all can flourish.

Such goals require developing a mindset very different from business as usual. This book supplies a roadmap to begin that transition. Digest it carefully, bearing in mind that human values transitions are more challenging than technical ones. Our technologies and techniques are beyond the dreams of our forefathers, and we are capable of devising many more. But our tough transition is fully grasping our symbiosis with our environment, repurposing our objectives for applying them. We work for more than our customers; we work for everything and everybody.

So sail on with your unending voyage of discovery. Help us leave our "global farm" in better shape than we found it. As the cases illustrate, once dedicated to the challenges, what we can do is amazing.

Robert W. "Doc" Hall
Chairman, Compression Institute

Acknowledgments

The authors would like to thank all organizations and individuals who have supported this project. Especially we would like to thank Steve Hope (Toyota Motor Europe), Carmel Mcquaid (Marks and Spencer), Andrew Mackenzie (Marks and Spencer), Louise Nicholls (Marks and Spencer) and Vidhura Ralapanawe (MAS). We would also like to acknowledge the support we have received from a number of brilliant researchers including Dr. Alastair Smith (Cardiff University), Dr. Ioannis Manikas (Greenwich University) and Dr. Evelyn Nava-Fischer (Cardiff University). We are indebted to all those who have endowed us with their ideas, encouragements and feedback throughout this project including a number of consultants at SA Partners. Last but not least, we are grateful to both Jonathon Porritt and Doc Hall who kindly took time out of their very busy schedules to read various drafts of this book and to write two valuable forewords as well as supplying us with their insightful comments.

All the proceeds of this book were dedicated to charitable organizations.

About the Authors

Keivan Zokaei is a lean thinker, management consultant and author. He is an honorary visiting professor at the University of Buckingham with years of experience at the cutting edge of lean thinking specializing in operations management and supply chain management. Previously Keivan was a director at the Lean Enterprise Research Centre at Cardiff University.

Keivan has also been an advisor to the auditor general in Wales on how to improve public services and has advised different European governments on the principles of lean and systems thinking as well as publishing a number of key reports commissioned by governmental or industrial bodies. He is a regular speaker at different international conferences and engages with both Lean professionals and general business executives through public media including appearances on BBC.

Hunter Lovins is president of Natural Capitalism Solutions. NCS helps companies, communities and countries implement more sustainable business practices profitably. Hunter has worked in countries from Afghanistan to New Zealand, and was recently asked by the Prime Minister of Bhutan to be an expert at a conference at a UN High-Level Meeting on transforming the global economy. She chaired the Civil Society Working Group charged with creating a global movement to enable Gross National Happiness to supplant GNP as the dominant paradigm.

Over her 30 years as a sustainability thought leader, Hunter has written hundreds of articles and 13 books. Her latest, *The Way Out: Kickstarting Capitalism to Save Our Economic Ass* (2012), is a sequel to her international best-selling, *Natural Capitalism*, now used in hundreds of college courses.

A founder of the field of sustainable management, Hunter has helped create several MBA programs and currently teaches sustainable business at Bainbridge Graduate Institute.

Hunter has won dozens of awards, including the European Sustainability Pioneer award, the Right livelihood Prize (the alternative Nobel) and this year the Rachel Carson Award. *Time Magazine* recognized her as a Millennium Hero for the Planet, and *Newsweek* called her the Green Business Icon.

Andy Wood joined Adnams in 1994 with responsibility for developing its customer service and supply chain operations. He joined the board in 2000 with the additional responsibility of sales, marketing and the wine business. He became managing director in 2006 and chief executive in 2010. Andy has an MBA, a doctorate from Cranfield University and in 2010 he was awarded an honorary doctorate from Anglia Ruskin University.

Andy is also chairman of the Local Enterprise Partnership for Norfolk and Suffolk and a director of Adnams Bio Energy an innovative joint venture between Adnams plc and the Bio Group where food waste is turned into renewable energy.

Prior to joining Adnams, Andy worked for Norwich Union, now Aviva, in a number of roles encompassing customer service, purchasing and change management.

Peter Hines was the co-founder of the Lean Enterprise Research Centre at Cardiff Business School. He holds a BA (MA) in geography from Cambridge University and an MBA and PhD from the University of Cardiff. Peter followed a successful career in distribution and manufacturing industry before joining Cardiff Business School in 1992. Since that time he initially led the Materials Management Unit and then chaired the Lean Enterprise Research Centre within Cardiff Business School.

He has undertaken extensive research into Lean thinking and supply chain management and has pioneered a number of key concepts, methods and applications in Europe and Australia, including supplier associations, value stream mapping and the Lean business system. He has written or co-written several books including *Creating World Class Suppliers* published by Pitman in 1994, *The Lean Enterprise* published by AMACOM in 1997, *Value Stream Management* published by Financial Times Prentice Hall in 2000 and *Staying Lean* first published in 2008 which won a Shingo Research and Professional Publication Prize Recipient for 2009.

Peter is chairman of S A Partners, a specialist consultancy organization that assists companies to apply Lean. S A Partners has grown rapidly to over

50 employees working with large multi-national firms such as Corus (Tata), Wartsila, Mars & Arla together with many smaller companies. It now has operating companies in the UK, USA, Ireland and Australia.

WHY LEAN AND GREEN? 1

Chapter 1

Introduction

Creating Leaner and Greener Companies

In 2003, a report published by the U.S. Environmental Protection Agency (EPA) said: "There is a strong and growing network of companies implementing, and organizations promoting, lean across the U.S. [and the rest of the world]. For those companies transitioning into a lean production environment, EPA has a key opportunity to influence their lean investments and implementation strategies by helping to explicitly establish with lean methods environmental performance considerations and opportunities. Similarly, EPA can build on the educational base of lean support organizations—non-profits, publishers, and consulting firms—to ensure they incorporate environmental considerations into their efforts. As several lean experts suggested, efforts to 'paint lean green' are not likely to get far with most lean practitioners and promoters. Instead, public environmental management agencies will be better served by being at the table with practitioners and promoters, seeking opportunities to fit environmental considerations and tools, where appropriate, into the context of operations-focused lean methods."[1] Nearly a decade later, unfortunately, this has largely not happened. We believe this is a missed opportunity, not just by governmental organizations, but more importantly by the business community. The economic recession is hardly an excuse and, in fact, should drive businesses to put resource efficiency and environmental protection on the top of their agenda.

Numerous companies around the world, large and small, have embarked upon some sort of continuous improvement journey, while few have begun to really see the opportunities that lie on the boundaries between economic continuous improvement and environmental betterment. In this book, we demonstrate how lean benefits "green intentions" by providing valuable experiences, techniques, and methodologies for engaging with the workforce, bringing about a cultural change, and using proven tools for measuring and improving systems holistically. Equally, we show how green can benefit the continuous improvement journey in an organization by acting as a powerhouse of innovative ideas, engaging people across the usual boundaries within the company and even beyond, and creating a positively charged environment to which everybody can relate.

There is a logical fit between economic and environmental waste reduction, in other words, between lean and green. We regard *lean* as a universal term for all forms of continuous improvement activities aimed at improving the economics of the firm, although readers will find a more elaborate description of lean thinking in Chapter 3. Similarly, we use the word *green* to capture all management practices aimed at improving the environmental performance of the firm. It is plausible that elimination of economic waste (lean) results in environmental waste reduction (green) and vice versa. However, this is not always the case. We have seen various cases where lean implementation has had no positive impact or even detrimental effects on the environment due to lack of integration of the green objectives. This is why lean cannot always be regarded to be intrinsically green2; there needs to be conscious design to realize the synergies between lean and green. Many well-known lean organizations that are widely acclaimed for their business excellence, among them Toyota, are simultaneously investing in improving their environmental performance by drawing upon their technical lean capabilities as well as their continuous improvement culture to push the boundaries of environmental management and resource efficiency.

Lean's greatest contribution to environmental management is in creating engagement and alignment. Lean offices across different industries have the potential to leverage a great deal of leadership and behavioral knowledge behind the implementation of the green objectives. This is about teaching green experts not *what to do*, but rather *how to get it done and then sustain it.*

Lean is about *tomorrow better than today*; it is about creating a learning organization to pursue systematic eradication of all types of constraints to the organization's economic goals. Lean firms see every problem as an opportunity to surface limitations to competitiveness. In doing so, they learn to create engaging environments or a *bottom-up organization* where every employee is empowered to bring about the necessary change and to sustain it. Environmental management, on the other hand, seems to be more concerned with technical fixes and top-down implementation. At least, so far. However, it must be noted that, across different industries, the current levels of lean maturity and business excellence are underpinned by several decades of continuous improvement and dissemination of knowledge from one area to another—something that green management does not benefit from yet. Over the last few decades, numerous approaches to continuous improvement have emerged such as Total Quality Management, lean thinking, Six Sigma, and the Theory of Constraints. On the other hand, the concepts of *green business management* are relatively newer to most industries and less widely adopted. Our environmental management toolkit is not nearly as mature or as well established as the economic "business excellence" toolkit. Moreover, the promoter communities such as authors, nonprofits, consultants, practitioners, and advisory bodies are much more established, more mature, and better equipped in the area of lean and economic continuous improvement than in green management.

In this book we explain how lean techniques and the culture that underpins lean can and should be translated for environmental improvements. We set off to show why the existing segregation between lean and green is unnecessary and rather arbitrary, demonstrating how leading firms use both lean and green as simultaneous sources of inspiration and innovation. Despite the lack of integration between lean and green so far, we believe this is a timely contribution evidenced by the increasing interest in the application of lean and continuous improvement approaches to the field of environmental management. There is already a body of knowledge emerging in this area alluding to the natural fit between business excellence and environmental management.[3,4]

Even more crucially, we aim to show a systematic way for organizations to design their green journey to eliminate green wastes and to generate green value. Currently, many organizations seem to be void of systematic methods for continuous green improvement. Environmental management systems and green interventions are by and large skewed in the direction of one-off technical fixes, end-of-pipe solutions involving a limited number of people and scattered improvements that hardly leave a lasting cultural change. On the other hand, the lean community has increasingly become aware of the importance of systematic improvements, paying attention to the cultural and leadership issues and orderly technical methods such as value stream mapping or road mapping techniques. We want companies to become green systematically and holistically by drawing on decades of experience accumulated by the lean community successfully designing countless continuous improvement journeys. As a result, we develop a body of evidence to demonstrate the efficacy of the application of the lean approaches and philosophies to green management while explaining the relationship between lean and green in a systematic rather than a piece-meal fashion. We demonstrate how the continuous improvement (CI) tools and techniques could be applied to environmental improvements. We have even created a suite of *lean and green* tools and techniques for those who like their lean toolkit. The tools and techniques introduced in this book are structured rigorously so that users will understand their applicability, strengths, and weaknesses in different contexts.

Several case studies and numerous examples are provided to illustrate the efficacy of continuous improvement approaches for environmental management. Each case study is used to shed light on a certain aspect of lean and green. These examples give practical guidelines on the application of continuous improvement concepts to environmental management, different approaches to lean and green, and the challenges and opportunities on the path to becoming lean and green.

Moving to a Leaner and Greener Economy

All the way, since the early nineteenth century, we have enjoyed unprecedented access to cheap—or even free—energy and resources leading us to disregard the real costs of our growth and prosperity. The twentieth century was only three years old when Henry Ford founded the Ford Motor Company and Orville Wright flew a plane for 12 seconds. By the end of the century, 58 million motor vehicles were manufactured per year and airplanes circled the globe. Nearly 30 million flights take off every year,[5] and only 4 nations emit more carbon than the United States' vehicle fleet. Give it another century, and it will all change at least as much again. But with the commodity and energy prices already at record levels, dwindling natural reserves, soaring demand from the developing world, and an array of looming sociopolitical disruptions, the days of cheap energy and resources are long gone. Only in the last two years, in the United Kingdom, the cost of filling up a Ford Mondeo with a 70-liter tank has soared from under £80 to £100. As we simply cannot sustain our existing levels of production and consumption, we need to take a new look at the entire economy—the way we design, produce, distribute, and even consume products and services. Focusing on desired outcomes rather than safeguarding last century's technologies offers opportunities for innovation that can simultaneously reduce our dependence on natural resources and enhance global prosperity.

The United States alone uses 25 percent of the world's oil, but represents only 5 percent of the global population. But how do Americans pay for it? They don't. In the words of Charles McMillion, president and chief economist of MBG Information Services in Washington in 2008, "it is remarkable that the United States still cannot competitively produce and sell enough to pay for imports and must continue borrowing almost $2 billion per day in today's global financial markets."[6] Interestingly, the money is borrowed from China and the sovereign wealth funds of the Middle East to buy imported oil, in competition with the Chinese, from the sovereign wealth funds of the Middle East. At the same time, the petroleum that fuels America is heavily subsidized. Most recent estimates show that the US federal subsidies to the oil industry were more than $502 billion between the years 2002 and 2008,[7] with an additional $16.8 billion going into producing corn-based ethanol, mostly for cars. Had this money been spent on public transit, more than twice as many jobs would have been created instead of putting the money into more fuel for cars.[8] This is certainly not "natural capitalism;" it is socialism for drivers. Seventy-five percent of Americans surveyed in 2007 by the National Association of Realtors felt that being smarter about development and improving public transportation are better long-term solutions for reducing congestion than building new roads.[9] In a study of 68 US cities in the state of Texas, it was estimated that the total congestion "cost" was $78 billion, representing 4.5 billion hours of delay and 6.8 billion gallons of excess fuel consumed.[10]

The opportunities in the nontransportation parts of the economy are arguably even larger. A report published by McKinsey and Company[11] illustrates that the energy-saving opportunities in the "stationary" sections of the US economy are worth $130 billion per year. The study shows that energy efficiency offers a vast, low-cost source of energy for the US economy in the coming years with excellent return on investment possibilities and with the potential to abate 1.1 gigatons of greenhouse gases annually. What we really require is true innovation in policy and technology. We need policy making that encourages entrepreneurial solutions that deliver and foster the convenience that our twenty-first-century life style demands, while also benefiting communities and climate.

Here is a story so crucial to all of us for creating a leaner and greener economy. In 2007, Toyota became the world's largest automobile manufacturer. Toyota swept to number one having set the aspirational goal of building a car that could go coast to coast, in the United States, on one tank of gas. In 2006, Toyota's profits were almost precisely Ford's losses. It is ironic that just as it was achieving world domination, Toyota began to falter and its shares slid to the lowest point since the depth of the recession. Some argued that her lean philosophy of innovation and quality fell victim to its relentless pursuit of growth. To beat Detroit, Toyota focused on building the big pickup trucks that conventional wisdom, and the marketing agencies, said Americans wanted. Some observers even commented that to sell more cars, Toyota had forgone quality in their flagship Prius and the smaller cars that built the brand, forcing the company to recall millions of vehicles to correct safety defects. Although the quality allegation were blown out of proportion, Toyota President and grandson of the founder, Mr. Akio Toyoda, acknowledged, "Quite frankly I fear the pace at which we have grown may have been too quick. We pursued growth over the speed at which we were able to develop our people and our organization, and we should sincerely be mindful of that. I regret that this has resulted in the safety issues described in the recalls we face today."[12] Add to this story the fact that, in 2009, General Motors, the icon of American business, entered bankruptcy, only to be bailed out by the US taxpayers. Unlike the European and Asian carmakers, that correctly identified small, fuel-efficient cars as the rising market, nearly every American automaker went down with their trucks and SUVs right up to the verge of bankruptcy.[13] Nevertheless, since then, GM seems to have changed course, beginning to invest in leaner and greener technologies. In November 2008, exactly when the US government was considering the bailout of the Big Three US automakers, Rick Wagoner, General Motor's then-chairman and chief executive officer (CEO), wrote for the *Wall Street Journal* exclaiming, "We have taken tough actions to cut costs, at the same time investing billions in fuel-efficient vehicles and new generations of advanced propulsion technologies."[14] He was right. At long last, the Detroit Big Three seemed to have begun waking up to the economic realities. The trio's productivity has improved, according to a leading industry benchmark; in 1995, the Harbour Group reported that a GM car on average took 46 hours to make, Chrysler 43, and Toyota only 29.4. Based on the latest available data, by 2007, GM had improved productivity to 32.3 hours per vehicle and Chrysler 30.4, while Toyota slightly slipped up to 29.9.[15] All evidence points to the fact that the Western manufacturers have worked to bridge the productivity and leanness gap with the Japanese. But that's not all. In a recent study, GM and Toyota jointly topped the list of the "greenest vehicles" of 2012 in the United States;[16] a clear sign of their recent efforts to make more fuel-efficient vehicles. It is not surprising that GM has gradually slid back into the black, beginning to recover from the shock of 2008. Becoming leaner and greener is only just beginning to save the American auto industry and can help save the rest of the—troubled—Western economy.

But we also need to mention that, at the same time the American auto manufacturers have been investing in bridging the lean and green gap with Toyota, Toyota herself has not stood still. Toyota—this icon of efficiency and effectiveness—showed remarkable perseverance, enduring the recent "hiccups" by going back to its core principles of respecting people and continuous improvement, as well as investing ever more ambitiously in green initiatives as a key differentiator for generations to come. Toyota, again, proved that they have the culture to turn a crisis into an opportunity. Through intensive self-reflection, they found many opportunities for improvement—problems that did not necessarily cause any recalls.[17] The allegations negative publicity regarding a supposed decline in product quality were in fact unfounded and Toyota was publically exonerated when NASA published a verdict, in February 2011, ruling out electronic technical faults causing sudden unintended acceleration in Toyota vehicles,[18] an issue that had sparked fashionable Toyota bashing attacks in the media. We will have a more elaborate look at Toyota in the following chapters to show how lean and green are part of the same continuous improvement mentality within Toyota.

Chapter-by-Chapter Overview

In the first chapter we explain what this book is trying to achieve, the importance of lean and green for individual companies, and the benefits it can bring to the economy as a whole. In the following chapter we describe the ever-growing need for leaner and greener business practices, especially during the economic down turn. We discuss how the economic downturn and the global environmental degradation are inextricably linked and why the solution for one is also the solution to the other. In the third chapter we take a deeper look at both lean and green, their meanings, importance, and synergies. The concept of lean thinking has been around for more than 20 years, and even before that, other similar continuous improvement concepts have been around for decades. Nonetheless, very few organizations have managed to create truly lean business systems such as the one of Toyota. Chapter 3 explores some of the less understood aspects of leanness and greenness and discusses their evolution independently as well as the enormous opportunities that exist in connecting them together. The first three chapters together make up the first section of the book and set the scene for Section 2, where we provide many more practical examples.

Section 2 illustrates how a lean and green business system can be created in practice and explains its benefits using case studies from different sectors of the industry from automotive to retail and from textile to brewing. In this section we put forward the lean and green business system model and elucidate its constituent components: lean and green process management, lean and green business leadership, lean and green strategy deployment, and lean and green supply chain collaboration. Each of the four chapters in Section 2 elaborates on an individual aspect of lean and green and provides practical case studies from leading organizations. Chapter 4 is titled and is about "Lean and Green Business Process Management." It looks at various cases, most notably Toyota; it also illustrates our suite of tools, which can be deployed in various situations to create leaner and greener processes. Chapter 5 then demonstrates the role of leadership by drawing on a unique success story: The story of Adnams Plc, a medium-sized brewery based in the UK that has turned a new page in addressing environmental issues through continuous innovation. Chapter 6 looks at the importance of strategy alignment and effective deployment of organizational objectives across different departments and levels in the company. In this chapter we provide a case study of Tesco, which has gone from being an average UK retailer to becoming the third largest global retailer in a fairly short space of time. Tesco has been successfully deploying their "steering wheel" as the method for strategy alignment for both economic and environmental objectives. Chapter 7 then, looks at lean and green across the extended supply chain. In that chapter we study Marks and Spencer, one of the greatest names in profitable greenness. Marks and Spencer have been investing in lean and green, not just in their own operations, but also across their end-to-end supply chains. Another case study, in Chapter 7, belongs to MAS—a leading textiles manufacturer in Sri Lanka. It supplies to Marks and Spencer, alongside many other leading lingerie brands such as Victoria's Secret and Nike. MAS has had a successful in-house lean or continuous improvement program for several years. They also drive their continuous improvement culture into environmental improvements from product design to process reengineering. The MAS case study is a clear example of how the principles of lean and green can be implemented anywhere in the world, even in developing economies, with tremendous results. All that is needed is a profound commitment to simultaneous lean and green improvement.

Section 3 draws some important conclusions and puts forward more practical advice for those who consider embarking upon their own lean and green journey.

Endnotes

1. Environmental Protection Agency 2003, pp. 3–4.
2. A. A. King and M. J. Lenox, "Lean and Green? An Empirical Examination of the Relationship between Lean Production and Environmental Performance," *Journal of Production and Operations Management* 10, no. 3 (2001): 244–256.
3. P. Hawken, A. B. Lovins, and L. H. Lovins, 1999. *Natural Capitalism*, (London: Earthscan Publications Ltd., 1999).
4. R. Hall, *Compression*, (New York: Productivity Press, 2009).
5. "OAG Reports 29.6 Million Flights Worldwide," Breaking Travel News, December 14, 2007, http://www.breakingtravelnews.com/news/article/btn20071214104331417/.
6. David Dickson, "Oil Woes Mask Improvement in Deficit, Demand for Imported Goods Slows; Exports Have Increased," *Washington Times*, August 17, 2008, http://www.washingtontimes.com/news/2008/aug/17/oil-costs-mask-improvement-in-deficit/
7. "Estimating U.S. Government Subsidies to Energy Sources: 2002–2008," Environmental Law Institute, http://www.elistore.org/Data/products/d19_07.pdf
8. Jane Holtz Kay, *Asphalt Nation: How the Automobile took over America and How We Can Take it Back*, University of California Press. Berkeley: CA. 1998 p. 129.
9. Nicholas Chang, "Survey Shows Americans Prefer to Spend More on Mass Transit and Highway Maintenance, Less on New Roads, Smart Growth America," October 25, 2007, http://www.smartgrowthamerica.org/narsgareport2007.html
10. "Transportation and Economic Prosperity," Surface Transportation Policy Project, n.d., http://www.transact.org/library/factsheets/prosperity.asp.
11. *Unlocking Energy Efficiency in the US Economy*, Report prepared by McKinsey and Company, July 2009.
12. Bryan Usrey, "Toyoda: Toyota Sacrificed Safety for Growth," *Automotive News Examiner*, February 23, 2010, http://www.examiner.com/x-32892-Akron-Sports-Car-Examiner~y2010m2d23-Toyoda-Toyota-sacrificed-safety-for-growth
13. "America's Other Auto Industry," *Wall Street Journal*, December 1, 2008, http://online.wsj.com/article/SB122809320261867867.html

14. R. Wagoner, "Why GM Deserves Support," *Wall Street Journal*, November 19, 2008, http://online.wsj.com/article/SB122705733362939557.html

15. "GM, Ford and Chrysler Strive to Become the Lean Three," GM, *Industry Week*, November 18, 2009, http://industryweek.com/articles/gm_ford_and_chrysler_ strive_to_become_the_lean_three_20441.aspx?Page=3?ShowAll=1.

16. "Toyota and GM Dominate List for Combined Fuel Economy, Industry Ratings," *Environmental Leader*, March 12, 2012, http://www.environmentalleader.com/2012/03/21/ toyota-and-gm-dominate-list-for-combined-fuel-economy-industry-ratings/.

17. Jeffrey K. Liker and Timothy N. Ogden, *"Toyota under Fire"* (New York: McGraw Hill, 2012).

18. The National Highway Traffic Safety Administration (NHTSA) office of the US Department of Transportation commissioned NASA to investigate the unintended sudden acceleration claims made against Toyota. NASA's report, published in February 2011, made it clear that NASA and NHTSA have not found any electronic or otherwise vehicle-based causes of large throttle opening that can result in unintended acceleration.

Chapter 2

Need for Lean and Green Business Practices in an Economic Downturn

Transforming Our Economy

Lean approaches have long been pursued because they save companies money. The same is true of green practices. Lean *and* green together are not only more effective, they are more profitable.

When those wild-eyed environmentalists at Goldman Sachs tell you that the companies that are the leaders in environmental, social, and good governance policies have outperformed the MSCI world index of stocks by 25% since 2005, with 72% of the companies on the list outperforming industry peers, there is a business case for sustainability.[1] Forty-five studies from the likes of the Economist Intelligence unit, McKinsey, AT Kearney, Deloitte, Harvard, MIT Sloan, and others show that companies that use leaner, greener practices and commit to such aspirational goals as zero waste, zero harmful emissions, and zero use of nonrenewable resources are outperforming their competitors.[2] More than that, 90% of CEOs surveyed by Accenture now believe that addressing sustainability will be key to their business success over the next decade.

Yet many companies, even those that have embraced lean practices, delay implementing more sustainable practices, saying that any investment that does not pay back in a few quarters will disadvantage them on Wall Street. This fixation on managing companies to suit a handful of Wall Street traders and the analysts who answer to them is harming both long-term shareholder value and the world.

Investments in more sustainable ways to run the business that cut costs for years to come, keep cash in the corporate coffers, and enable higher dividends should be made regardless of what stock analysts think. Traders benefit whether the market goes up or down. Companies and their employees, managers, and owners do not. The communities in which they do business do not benefit when managers lay off employees to cut costs before quarterly reports are due. The pursuit of short-term profits, and the failure to manage a company for the long term, has brought the world to financial collapse, and worse, to the brink of environmental collapse.[3]

Enter Paul Polman, CEO of Unilever. Polman rejects Milton Friedman's optimization of short-term profits, stating, "Our new business model will decouple growth from environmental impact. We will double in size, but reduce our overall effect on the environment. Consumers are asking for it, but governments are incapable of delivering it. It is needed for society and it energizes our people— it reduces costs and increases innovation."[4]

In early 2011 Polman stated:

> The focus on delivering short-term shareholder value has led to widespread addiction to quick artificial highs—rather like a junkie hooked on heroin or a financial trader on cocaine. The ultimate cost of short-termism was the financial crisis of 2008–09. ... Too many investors have become short-term gamblers: the more fluctuations in share price they can engineer, the better it is for them. It is not good for the companies or for society, but it is influencing the way firms are being run. ... To drag the world back to sanity, we need to know why we are here. The answer is: for consumers, not shareholders. If we are in synch with consumer needs and the environment in which we operate, and take responsibility for society as well as for our employees, then the shareholder will also be rewarded.[5]

In October 2011, Polman announced that Unilever would not submit quarterly reports. The company's share price fell 10%. His answer: good, that's not the sort of investor we want. His plan to double sales *and* halve the environmental impact of Unilever's products over the next 10 years, improve the nutritional quality of its food products, and link half a million smallholder farmers and small-scale distributors in developing countries to its supply chain will require the sort of longer-term investments that quarterly analysis would penalize. Polman stated, "The Occupy Wall Street movement sends out a very clear signal. If you look out five or ten years … consumers will not give us a sense of legitimacy if they believe the system is unfair or unjust. Companies that miss the standards of acceptable behaviour to consumers will be selected out."[6] Polman understands that business as usual is a recipe for erosion of value in the near term and global disaster in the longer run.

He's right. The global economy rests on a knife-edge. The financial crash of 2008 caused $50 trillion dollars and 80 million jobs to evaporate.[7] And the wreck is not over. The International Labour Organisation (2011) sets forth the grim statistics:[8]

- Sixty-nine of 118 countries with available data show an increase in the percentage of people reporting worsening living standards in 2010 compared to 2006.
- People in half of 99 countries surveyed say they have little confidence in their national governments.
- In 2010, more than 50% of people in developed countries lack decent jobs (in Greece, Italy, Portugal, Slovenia, and Spain, it's more than 70%).
- The share of profit in GDP [gross domestic product] increased in 83% of countries studied between 2000 and 2009, but productive investment stagnated globally during the same period.
- Growth in corporate profits increased dividend payouts (from 29% of profits in 2000 to 36% in 2009) and financial investment (from 81.2% of GDP in 1995 to 132.2% in 2007). Bankers regained their bonuses, but workers face falling wages.
- Food price volatility doubled during 2006–2010 from the prior five years. Financial investors benefit from this; food producers do not. Remember it was a food riot that touched off the Arab Spring in Tunisia.

The current economic crisis is different from prior market downturns, however. In 2009, Jonathan Porritt, adviser to the Prince of Wales, and writer of the Foreword to this book warned, "People seem blind to the fact that the causes of the economic collapse are exactly the same as those behind today's ecological crisis—and behind accelerating climate change in particular." Porritt wrote in support of the UK government's chief scientific adviser, Sir John Beddington, who predicted, "A 'perfect storm' of food shortages, scarce water and high-cost energy will hit the global economy before 2030." Porritt warned, "There is a simple conclusion here: the self-same abuses of debt-driven 'casino capitalism' that have caused the global economy to collapse are what lie behind the impending collapse of the life-support systems on which we all ultimately depend." He believes that the storm will hit by 2020.[9]

New York Times columnist Tom Friedman (2009) agrees, "Let's today step out of the normal boundaries of analysis of our economic crisis and ask a radical question: What if the crisis of 2008 represents something much more fundamental than a deep recession? What if it's telling us that the whole growth model we created over the last 50 years is simply unsustainable economically and ecologically and that 2008 was when we hit the wall—when Mother Nature and the market both said: 'No more.'"[10]

In 2010, Global Biodiversity Outlook 3,[11] building on the 2005 Millennium Ecosystem Assessment,[12] warned that climate change and other assaults have put all of the Earth's major ecosystems at risk; three are tipping into collapse: by the end of this century, business as usual, there will be no living coral reefs on planet earth, the Amazon now releases more carbon than it soaks up, and the acidifying oceans risk ending life as we know it.[13]

It is common for business leaders to greet such apocalyptic scenarios with denial. Didn't that old book *Limits to Growth* forecast such catastrophes? And it hasn't happened yet, has it? A recent analysis of the projections initially set forth in the 1972 book tracks the grim forecasts of that model against such observed trends as constraints of nonrenewable resources remaining, population growth, food per capita, industrial output per capita, and so on, over the period 1970–2000. It finds them precisely on target for the predicted collapse in 2030.[14]

Sobering.

The situation is already dire for vulnerable populations.[15] While the richest 7% (about half a billion people) spew out 50% of global carbon emissions, the poorest three billion emit only 7%.[16] But already millions of drought-driven refugees roam the horn of Africa and the Sahel.[17] Millions more have been displaced by record flooding in Pakistan, even as additional billions are threatened by failing glacial melt water from the Himalaya to Columbia.[18]

Let us assume, however, that none of this is true: climate change is a hoax, and all the global weirding is caused by sunspots or microwaves. Don't go to the casino on the odds of that being true, but in a way, it doesn't matter. Hundreds of companies and entrepreneurs are showing that even if you do not believe any of the science on the climate crisis, resource constraints, or environmental challenges, the best way to profit is to behave precisely as you would if your senior management were tree huggers.

DuPont was one of the early examples of this; about a decade ago, the company's leaders pledged to cut its carbon emissions 65% below their 1990 levels, and to do it by 2010.

That's a bit more ambitious than the United States, which still refuses to ratify the Kyoto protocol agreeing to cut emissions 7% below 1990 levels by 2012. Did DuPont join Greenpeace? No. The company made its announcement in the name of increasing shareholder value. And it delivered on that promise. The value of DuPont stock increased 340% while the company reduced global emissions 80% below 1990 levels, by 2007. Doing this created a financial savings for the company of $3 billion between 2000 and 2005.[19] It cost less to implement energy savings measures than it did to buy and burn fuel. In short, DuPont was solving the climate crisis at a profit, estimating in 1999 that every ton of carbon it no longer emitted saved its shareholders $6.00. By 2007, DuPont's efforts to squeeze out waste were saving the company $2.2 billion a year. The company's profits that year? $2.2 billion.[20]

The world's largest retailer found the same advantages from setting and delivering on substantial sustainability commitments. Like Marks and Spencer, Ikea, Tesco, and Sainsbury's, Walmart is investing in green stores and supply chain reforms.[21] When Walmart asked its global supply chain of 100,000 companies to measure their carbon footprints and report them to the Carbon Disclosure Project,[22] it even drove sustainability into small companies. In its 2012 sustainability report, Walmart described how it is getting 22% of its electricity from renewable sources including onsite solar panels, fuel cells, and micro-wind turbines. Walmart has an additional 180 renewable energy projects under development.

Walmart has doubled sales of locally grown food, while saving its customers over $1 billion dollars. It has committed to source $20 billion from women-owned businesses in the United States over the next 5 years. The company's Zero Waste Program claims that 80.9% of waste generated in its US operations has been diverted from landfills, cutting carbon emissions the equivalent of 11.8 million metric tons of carbon dioxide. Its UK subsidiary, ASDA, sends zero waste to landfill. Walmart's stores in China and Brazil have cut waste to landfill in half.[23] Not surprisingly, such programs have enabled Walmart to cut the percentage of people who have an unfavorable view of the company from 38% in 2005,[24] when Walmart's sustainability program was announced, to 20% in 2010.[25]

Existing companies have huge opportunities to "intrapreneur" sustainable solutions.[26] In May 2005, Jeffrey Immelt, who replaced Jack Welch as the CEO of General Electric, announced "ecomagination."[27] Immelt, seeking to restore a culture of innovation at GE, committed to double the company's investment in environmental technologies to $1.5 billion by 2010. GE would reduce the company's greenhouse-gas emissions 1% by 2012 (without action, emissions would have risen 40%). Immelt stated, "We believe we can help improve the environment and make money doing it."[28] Critics charged that GE was *greenwashing*, simply branding some of its existing products as green and changing little else. Hypocrisy, however, is often the first step to real change. Immelt worked with Green Order to map out a strategy to deliver on his promise of green innovation.

A little less than a year after the announcement of ecomagination, Immelt revealed that his green-badged products had doubled in sales over the prior two years, with back orders for $50 billion more, significantly exceeding the initial prediction of $12 billion in sales by 2010. Over the same time frame, sales of the rest of GE products increased only 20%. GE also found that in 2006 it had reduced greenhouse gas emissions by 4%, easily beating its 2012 target of one percent. GE worked with Walmart to commercialize more cost-effective LED light bulbs and other efficiency technologies. Because of such success, GE committed $10 billion more to ecomagination research and development to grow its portfolio of environmentally sensitive products, services, and technologies.

Even in a down economy, ecomagination revenues rose from $5 billion in 2005 to over $25 billion in 2010. The program enabled GE to cut its emissions by 22% in 2009 compared to its initial goal of 1% in 2004. By 2015, GE expects to cut the energy intensity of its operations by 50%.[29] In his annual letter to shareholders, Immelt reported, "ecomagination is one of our most successful cross-company business initiatives. If counted separately, 2009 ecomagination revenues would equal that of a Fortune 130 company and ecomagination revenue growth equals almost two times the company average."[30]

Does this make Walmart and GE sustainable companies? Of course not. But programs to wring waste from supply chains and implement more sustainable practices are clear indicators that even the largest companies recognize that there is a business case for moving in this direction. *Harvard Business Review* summed it up recently, "Sustainability isn't the burden on bottom lines that many executives believe it to be … sustainability should be a touchstone for all innovation. … In the future, only companies that make sustainability a goal will achieve competitive advantage."[31] A 2010 study by Accenture found that 93% of CEOs surveyed agree, admitting that "sustainability will be critical to the future success of their companies, and could be fully embedded into core business within ten years."[32]

Yet enormous savings remain to be captured. Natural Capitalism Solutions (NCS)[33] and other sustainability consultants work with such companies to help them cut waste, transform how they make products, and implement more sustainable ways of doing business. One company with which NCS worked had a practice of leaving its 6,300 computers and monitors turned on 24/7. Various urban myths about shortening the life of the computer by turning it off, or claims that the IT department required them to be left on, led the company to waste energy and money. NCS consultants pointed out that simply posting a policy that employees turn computers off when no one is in front of them would save $700,000 the first year.[34] In the United States alone, $2.8 billion a year is wasted simply because computers are left on when no one is using them. Such IT costs can represent a quarter of the cost of running a modern office building.[35] In March 2010, Ford Motor Company followed suit, announcing that it had saved $1 million that year by shutting off unused computers.[36]

A team from NCS also worked with a large distribution center, a 7 million-square-foot warehouse, in which 500-Watt light bulbs shone down on the tops of boxes stacked floor to ceiling. The workers below used task lighting so they could see where they were going. Simply flipping a switch would save $650,000 dollars a year.[37] Nationally the United States wastes $5 to 10 billion each year leaving unnecessary or redundant lights on.[38]

The savings from eliminating this waste are free—or better than free. And they exist throughout American businesses. It should come as little surprise that in 2008 American businesses used twice as much energy to produce a unit of gross national product (GNP) as their competitors in Europe and Asia, where adherence to the Kyoto Protocol is driving competitiveness. The American Council for an Energy Efficient Economy estimates that the US economy wastes 87% of the energy it consumes.[39] Far from constraining innovation, signatories to the Kyoto Protocol, which obliges them to save energy to cut carbon emissions, have driven their innovation, saving money in the process.

In 2012, as companies struggle to find ways to increase the productivity of their workers, the focus has shifted to employee engagement; 85% of senior managers surveyed believe that managing human capital is the most important means of improving productivity.[40] Companies with engaged workforces achieve higher earnings than organizations that fail to engage their employees. A recent Gallup poll found that, "Engaged organizations have 3.9 times the earnings per share growth rate compared to organizations with lower engagement in the same industry."[41] Gallup reiterates that there are ways to capture employees' interests, but an array of studies show that engaging employees on sustainability turns out to be a powerful motivating tool for delivering improved customer satisfaction, increased productivity, and reduced employee turnover and absenteeism. Americans feel worse about their jobs than ever before,[42] a situation that is costing $300 billion a year in lost productivity.[43] Coupling of sustainability training and employee engagement is a particularly effective way to enhance productivity, retain customers, and motivate a workforce. Those skilled in lean or other continuous improvement approaches in an organization likely have years of experience on how to engage the workforce to sustain improvements. Therefore, one of the key contributions of lean to green can be in creating bottom-up engagement. Greening your business engages the workforce even further. The two are mutually reinforcing.

Firms seeking competitive advantage today find this an especially potent mix. Ninety-two percent of the so-called Millennials say that they want to work for a responsible company.[44] *Fortune* magazine reports that Generation Y wants to be part of a workforce that has a positive impact on the world.[45] Replacing a good employee costs companies between 70 and 200% of the employee's annual salary.[46] Companies that want to attract and retain the best talent of the new generation will ensure that their employees understand and are engaged in the organization's commitment to sustainability.

Enabling employees to become more engaged with sustainability and better able to communicate company values to customers can be the defining factor between success and leaving money on the table. But traditional, class-based "Sustainability 101" approaches are expensive and not very effective ways to motivate workforces. Smart companies are using interactive e-learning modules that tie the concepts directly to what a person does at work. The platform enables companies to update online learning modules as new content becomes available. Assessment and incentive programs track employees' progress through the educational material.

The profitable efforts described previously are important first steps in tackling the economic and ecological challenges facing us, but they are wholly insufficient. Although it is increasingly commonplace for companies to tout their green credentials by engaging in cost-cutting measures they should be doing for basic business reasons, the world remains on a crash course to disaster.

In March 2012, the Organisation for Economic Co-operation and Development (OECD) released a study warning that unless world leaders take immediate and coordinated action, modern industry will lock the world into a calamitous temperature rise of up to 6 degrees. In such a world, demand for energy, food, and water will overwhelm the planet.[47] In April, the International Energy Agency (IEA) agreed, warning that the governments of the world are failing to act in ways that would prevent catastrophic climate change, locking the world into an inevitable 6-degree warming. "To meet the carbon cuts that scientists calculate are needed by 2020," the IEA says, "the world needs to generate 28% of its electricity from renewable sources and 47% by 2035. Yet renewables now make up just 16% of global electricity supply."[48] In 2012 and 2013, the WorldBank and the International Monetary Fund (IMF) released similar warning.

The world needs corporate leaders, too, to shift from incremental efficiency measures to transformative innovation in energy and in all of the ways that companies use resources. The reason need not be, as some snarky articles have alleged, CEOs undertaking charity by shouldering social responsibility.[49] The reality is quite the contrary. Ernst & Young's 2010 Business Risk Report concluded, "In order to maintain their corporate image and reduce environmental impact, companies must take proactive measures, including more complex decisions regarding capital spending, production procedures, and installed technologies." The report found that, despite uncertainties in the regulatory environment, companies must prepare for changes in regulation and carbon trading schemes. It placed what it called "radical greening" and "the need for social acceptance and corporate social responsibility" in the top ten risks facing business. "The risk will rise again in the future," the report concluded, stating, "successful companies will be those who put environmental policy at the top of their agenda and adapt their business to that goal. A consumer products commentator argued, 'As growth resumes and environmental degradation continues this will re-emerge as a very powerful force in shaping business.'"[50]

A new approach by Blue Earth Network is showing companies how to prosper by solving the world's toughest problems. This network of some of the best American sustainability consulting firms helps organizations build the sort of iconic brands that give companies such as Apple their fanatically loyal customers. It helps companies discover how behaving more sustainably can better deliver their true purpose, and innovate solutions to the real needs of their customers, achieving greater profitability.[51]

This approach integrates the Natural Capitalist approach that the innovative companies portrayed in this book have followed. Notice that as companies *implement* lean management and sustainability, they follow the first principle of Natural Capitalism: buying time by using all resources, anything taken from the Earth or borrowed from future generations, dramatically more productively. This is profitable now and solves or delays many of the most pressing challenges.[52] Many of the companies that claim green credentials are doing this. That is good, and the lean literature is full of success stories of enhanced productivity, but efficiency alone is insufficient to deliver durable competitive advantage, or to solve the gnarly challenges facing business and the world.

Blue Earth Network helps companies implement the second principle: redesigning how they make and deliver their products and services using such authentically more sustainable approaches as Walter Stahel's Cradle-to-Cradle concepts,[53] Biomimicry,[54] and the circular economy.[55] Nature makes a wide array of products and services that run on sunlight, producing neither waste nor toxics. The design of macroeconomic systems and microeconomic enterprises should mimic healthy, native ecosystems in diversity, adaptability, resilience, and local self-reliance. As Biomimicry founder Janine Benyus says, "After 3.8 billion years of research and development, failures are fossils, and what surrounds us is the secret to survival. The more our world looks and functions like this natural world, the more likely we are to be accepted on this home that is ours, but not ours alone."[56]

Companies such a Calera[57] are using seawater and 92% of the carbon dioxide waste from the Moss Landing, California, power plant to create cement in the same way that sea creatures create their calcium silicate shells. Every ton of cement the process makes sequesters half a ton of CO_2, in just the way that coral reefs are formed. Investors include the noted green venture capitalist Vinod Khosla, as well as such traditional polluters as Peabody Coal.

Recognizing that green plants do not see CO_2 as the biggest poison of our time but rather use it to create starches and glucose, the building blocks of life, scientists such as Dr. Geoff Coates at Cornell are mimicking this process, using CO_2 and catalysts to make biopolycarbonates, a biodegradable plastic that is almost 50% CO_2 by weight. "It's highly abundant and really cheap," says Dr. Coates. He is using similar catalysts to create Styrofoam from orange peels.[58] These design approaches are more profitable and create manufacturing that is locally based, job creating, and durable.

The third principle is to manage all institutions to be restorative of human and natural capital. These forms of capital, unlike the financial and manufactured capital in whose increased velocity of trade Wall Street delights, are not enhanced by globalization. It may be true that increasing the speed of trading stuff and money increases the wealth of some, but it does so at the cost of the environment and human community. As the various green GDP projects warn, this is not a net increase in wealth and well-being—it is only a conversion of some forms of wealth from others. The place-based forms of capital are not counted on any balance sheet (though the Puma Environmental return on investment [ROI] is a pioneering effort to correct this lapse), but they are the forms of capital without which no economy can exist. Their loss impoverishes us all. Managing companies in ways that are unconscious of this has brought the world to the verge of collapse. The Third Global Biodiversity Outlook warns that if ecosystems don't survive, neither will businesses. As Ray Anderson, the business leader who chaired the President's Council on Sustainable Development bluntly asks: "What's the business case for ending life on earth?"[59]

Making truly sustainable management as profitable as mere efficiency will require some changes in how business is conducted. A good start would be to implement the basic market principle of full cost accounting, or as perhaps better described, honest accounting.[60] Business practices that do not tell the ecological truth, that externalize environmental and social costs, and that drive companies and all of us to exceed the carrying capacity of the ecosystems to support life (and, lest the obvious be missed, necessarily all economic activity) need to be recognized as bad and unethical business.[61] Accounting arose when managers realized that if they did not have honest information about the financial status of their company they could not manage intelligently. Today, an equally momentous transition is needed in how businesses and governments keep accounts. The recognition of this has led to the creation of innovative business schools now teaching a holistic management philosophy to a whole new generation of entrepreneurs and corporate leaders.[62]

Sustainability advocates have urged companies to manage according to a *triple bottom line*, to achieve profit, but also to protect people and planet. While a tempting formulation, it has, in practice, led to managers bolting programs promoting the environment and social well-being onto balance sheets as cost centers, reducing the traditional measure of profit. Not surprisingly, this is a hard sell to bottom-line focused managers.

A more useful approach is the Integrated Bottom Line (IBL).[63] Attracting the interest of the Institute of Chartered Accountants of England and Wales,[64] and Harvard Professor Eccles' *One Report,*[65] this approach recognizes that profit is a valid metric, but only one of many criteria that give a company enduring value. As the examples in this book show, a corporate commitment to more sustainable practices:

- raises profitability by cutting energy and materials costs in industrial processes;
- drives innovation, leading to increased top-line revenues;
- supports better facilities design and management, and fleet management to increase effectiveness;
- delivers sector leadership and first-mover advantage;
- reduces risk and unbooked legal liabilities, protecting companies' franchise to operate;
- provides greater access to capital;
- improves corporate governance;
- enhances core business value through better government relations;
- builds brand equity by differentiating product and enhancing reputation;
- enhances competitive advantage through increased market share;
- underpins a company's ability to attract and retain the best talent;
- increases employee productivity and health;
- improves communication, creativity, and morale in the workplace; and
- builds better supply chain and stakeholder relations.

In 2012, the International Integrated Reporting Council began work on developing just such protocols. In the U.S. the Sustainability Accounting Standards Board began similar work. Both of these, if successful, will transform accountancy, and give business managers far more accurate metrics by which to drive profitability.

Taken together, these advantages confer the ability to be "first to the future." Well-managed companies that practice this approach will be the billionaires of tomorrow. Indeed, some financial advisors now state that a commitment to sustainability is the hallmark of good corporate governance and one of the best indicators of management capacity to protect shareholder value.[66]

Leading economists are beginning to think similarly, asking whether humans exist to serve economic goals, or whether the economy should be designed to serve humanity. It is not an easy issue to grapple with. Western economies are now 70% dependent on consumer spending, which is why after the 9–11 attacks President Bush implored Americans to "go shopping" to restore the American economy. (Nobel economist Paul Krugman, 2006, wondered whether there might be a nobler way for Americans to show support for their country.[67]) It also explains why current economic policy is aimed more at ensuring the health of banks and Wall Street than preserving intact ecosystems and human communities, although the Occupy movement and others have returned the question of the true purpose of the economy to the fore.

The crux of the question is, what is it that we wish to achieve? Measures like GNP claim to answer this. We are expected to be happy when it grows, and worried when it falls. But GNP is actually a very strange measure of anything. It counts only the velocity of the flow of money and stuff through the economy as they change hands in economic transactions. The more money that gets spent, conventional wisdom says, the better off we are. But are we? If you volunteer at a home for the elderly, you've done nothing to increase the GNP. A divorcing cancer patient who gets in a car wreck adds handsomely to the GNP as money goes for insurance, repairs, and medical bills. But is she any better off? Clearly not.

GNP's usefulness has been questioned since it was invented in the 1930s. Simon Kuznets, one of the architects of this way of keeping national systems of account, warned, "A nation's welfare can scarcely be inferred from their national income."[68] On March 18, 1968, Robert Kennedy put it more elegantly, stating:

> Too much and for too long, we seemed to have surrendered personal excellence and community values in the mere accumulation of material things. … That gross national product counts air pollution and cigarette advertising, and ambulances to clear our highways of carnage. It counts special locks for our doors and the jails for the people who break them. It counts the destruction of the redwood and the loss of our natural wonder in chaotic sprawl. … Yet the gross national product does not allow for the health of our children, the quality of their education or the joy of their play. It does not include the beauty of our poetry or the strength of our marriages, the intelligence of our public debate or the integrity of our public officials. It measures neither our wit nor our courage, neither our wisdom nor our learning, neither our compassion nor our devotion to our country, it measures everything in short, except that which makes life worthwhile.[69]

Is the point of the economy to enrich the 20% of the population that owns 80% of global wealth, at the cost of misery for the rest? Shouldn't good capitalists seek to enhance and productively use all forms of capital, human and natural, as well as manufactured and financial, to increase their ability to durably create more wealth?

This concept seemed quaint to most economists until 2009 when Nobel Prize Laureates Dr. Joseph Stiglitz, Dr. Amartya Sen, and French economist Jean-Paul Fitoussi released the results of the two-year Quality of Life

Commission they chaired.[70] Convened by the former French President Nicholas Sarkozy, the commission brought together economists and keepers of national accounts from across the world. They concluded that assessing a population's quality of life would require metrics from at least seven categories: health, education, environment, employment, material well-being, interpersonal connectedness, and political engagement. They also decided that any nation serious about progress should start measuring its *equity*— that is, the distribution of material wealth and other social goods—as well as its economic and environmental sustainability.

In the wake of the report's release, the European Union's statistical office and the OECD began debating how to implement more representative indicators of genuine well-being and to move the focus away from GDP. "We want policies that reflect our values, but nobody says what those values are," Stiglitz told a *New York Times* reporter. "The opportunity to choose a new set of indicators is tantamount to saying that we should not only have a conversation about recasting GDP. We should also, in the aftermath of an extraordinary economic collapse, talk about what the goals of a society really are."[71]

British statistician Nic Marks (2010) explores what it is that makes life worthwhile, noting that when people are actually surveyed and asked this question they say that they most want happiness, then love, then health, and only belatedly do they list getting more money.[72] If these are the natural human aspirations, Marks asks, then "Why are statisticians not measuring that? Why are we not thinking of the progress of nations in these terms, instead of just how much stuff we have?" He cites the Happy Planet Index, created by the New Economics Foundation,[73] which plots the *happy life years* of a country's citizens against the ecological footprint of that country. For example, the United States and the Persian Gulf states in the Middle East both use a lot of resources to deliver quality of life to their citizens. Well, most of them. In the wake of the 2008 collapse, poverty and unhappiness are rising in most Western nations. In contrast, countries such as Afghanistan and those in Sub-Saharan Africa use relatively few resources, but condemn their people to an average life span of 45 hard years. Marks cites an evidence-based study exploring what enables people to be happy. The answers: social relationships, being active, taking notice of one's surroundings, life-long learning and curiosity, and the quintessentially uneconomic activity of giving. All are activities that have little cost to the climate or the ecological integrity of the planet. "Happiness," concludes Marks, "Does not cost the earth."[74]

Some countries have already shifted their approach to national accounts. The concept that the Stiglitz commission was debating has come to be called Gross National Happiness (GNH).[75] First put forth in 1972 by the fourth ruler of the tiny Himalayan Kingdom of Bhutan, it codifies the practice of Buddhist economics profiled in E. F. Schumacher's landmark book, *Small Is Beautiful*.[76] Faced with an economy that was stagnating as it sought to pursue Western ways to enhance productivity, the King of Bhutan decided to focus instead on providing sustainable economic development, preserving and promoting cultural values, conserving the environment, and practicing good governance.

Dasho Karma Ura, president of the Center for Bhutan Studies in the capital Thimphu notes, "Defining happiness is not what is important. What is important is providing the conditions through which people can achieve happiness as they understand it."[77] GNH guidelines are now being used in Costa Rica and Chile, and are being adopted in Brazil, India, Haiti, and France. The Canadian Index on Well-being, the Measuring Australia's Well-being project, the State of the USA Index,[78] and the Living Planet Index of the USA are each trying to measure the genuine well-being of their countries' citizens.[79] The growing discipline of community indicators is a local version of the same approach.[80]

Bhutan and this growing number of countries have recognized that the developing world cannot lift itself from poverty in the same inefficient way that the West has done. The prime minister, Jigme Thinley stated, "Happiness is very serious business. The dogma of limitless productivity and growth in a finite world is unsustainable and unfair for future generations."[81] On the other hand, there are many desirable outcomes in which there can be extensive growth—community well-being, education, knowledge, music, culture, happiness, and so on. The things that cannot grow indefinitely are the use of finite, nonrenewable resources or the emissions of greenhouse gasses.[82]

In April 2012, Thinley chaired a high-level meeting at the United Nations, attended by over 600 experts, leaders of civil society, religious leaders, and business people. It undertook to transform the existing economic paradigm and reframe the UN's Millennium Development goals around the achievement of GNH. Bhutan has committed to staff a global secretariat for Gross National Happiness and to convene in 2014 a Bhutan Woods conference to reframe the now clearly failing economic institutions created by the world powers at Bretton Woods after World War II.[83]

What does this all have to do with business?

A lot. Bhutan has commissioned an International Expert Working Group, chaired by Dr. Robert Costanza, to lay the intellectual architecture of a new development paradigm, based on the concept of GNH.

A new report describes how switching to the use of the GNH concept would net the European manufacturing sector $630 billion by 2025.[84] A growing population has a lot of needs. Meeting those needs in sustainable ways is a business with a long future. The legacy industries of agriculture, car manufacturing, energy production, water supply, and all of the other outmoded bits of our economy will all be transformed. Yes, this is a formidable task, but it is precisely the sort of investment that will lift economies from recession and build the basis for future prosperity.

The green economy is already emerging at the local level, as people realize that they must build their own resilience to an unaccountable global economy.[85] The failure of global leadership to agree to climate protection at Copenhagen, Cancun, Doha, and Durban spurred cities to implement sustainable practices. Denmark's Samso Island is 100% renewably powered.[86] The German town Wildpoldsried is as well, producing 321% more energy than it uses and selling the excess for $5.7 million each year.[87] Over half of Germany's renewables are owned not by utilities but by farmers and citizens.[88] San Francisco is on track to be 100% renewable by 2020, Scotland by 2020, Germany by 2050.

National policies like Germany's Feed-in-Tariff (FiT) have unleashed Germany's renewable energy industry, underpinning German prosperity. In their first four years, FiTs created 480,000 new jobs and cut the unit cost of solar panels enough to reach grid parity (costing the same as grid electricity) by 2013.[89] The program added only two to three Euros per month to electricity bills in Germany, roughly $50 to customers' electricity bills each year, for a total of €8.6 billion. Deutsche Bank found that far from costing the economy, the savings outstripped the total cost of payments made by households. Had customers bought electricity from conventional coal generation, Germans would have paid an additional €9.4 billion.[90] Bill Clinton said if the United States implemented a similar program it would create 2.5 million jobs.[91]

As Achim Steiner, the brilliant undersecretary general and executive director of the UN Environment Programme says:

> The financial, fuel and food crises of 2008 are in part a result of speculation and a failure of governments to intelligently manage and focus markets. But they are also part of a wider market failure triggering ever deeper and disturbing losses of natural capital and nature-based assets coupled with an over-reliance of finite, often subsidized fossil fuels. The flip side of the coin is the enormous

economic, social and environmental benefits likely to arise from combating climate change and re-investing in natural infrastructure—benefits ranging from new green jobs in clean tech and clean energy businesses up to ones in sustainable agriculture and conservation-based enterprises.[92]

Joseph Stiglitz (2008) agrees, pointing out, "Spending money on needed investments—infrastructure, education, technology—will yield double dividends. It will increase incomes today while laying the foundations for future employment and economic growth. Investments in energy efficiency will pay triple dividends—yielding environmental benefits in addition to the short- and long-run economic benefits."[93]

This is good news, as already half of the world's people live in cities, three quarters by 2050. Projections warn that in the next decade or so, China will seek to move into cities, yet unbuilt, more people than there are in the United States. Just the copper wire this would require is more than current world copper production.[94] Business as usual, by 2030 China will want more oil than the world now lifts.[95]

The world as we know it stands poised between unprecedented prosperity and collapse. The decisions you make in your business and in your personal life will determine which it shall be.

Welcome to the future.

Endnotes

1. Margo Alderton, "Recent Report Finds Corporations that Lead in Corporate Responsibility Also Lead in the Market," Socially Responsible Investing 07-11 17:57, http:www.croassociation.org/?g=node/490/$soesg=printime. See also: http://www.natcapsolutions.org/business-case/GoldmanSachsReport_v2007.pdf.
2. For a list of these studies, see Natural Capitalism Solutions, http://www.natcapsolutions.org.
3. "UN Global Compact, Accenture Release Findings of Largest CEO Research Study on Corporate Sustainability," United Nations Global Compact, June 22, 2010, http://www.unglobalcompact.org/news/42-06-22-2010.
4. Andrew Saunders, "The MT Interview: Paul Polman of Unilever," *Management Today*, March 1, 2011, http://www.managementtoday.co.uk/features/1055793/MT-Interview-Paul-Polman-Unilever/
5. Ibid.

6. Jo Confine, "Paul Polman: 'The power is in the hands of the consumers,'" *The Guardian*, November 21, 2012, http://www.guardian.co.uk/sustainable-business/unilever-ceo-paul-polman-interview

7. Pia Lee-Brago, "ILO: Recession, Massive Job Loss Threaten Global Economy," *Philippine Star*, November 2, 2011, http://www.philstar.com/Article.aspx?articleId=743718&publicationSubCategoryId=.International Labour Organization World of Work Report 2011, Third World Network, October 2011, http://www.twnside.org.sg/title2/resurgence/2011/254/econ2.htm.

8. International Labour Organization World of Work Report 2011, Third World Network, October 2011, http://www.wnside.org.sg/title2/resurgence/2011/254/econ2.htm

9. Jonathan Porritt, "Perfect Storm of Environmental and Economic Collapse Closer Than You Think," *The Guardian*, March 23, 2009.

10. Tom Friedman, "The Inflection Is Near?" *New York Times*, March 7, 2009, http://www.nytimes.com/2009/03/08/opinion/08friedman.html.

11. Global Biodiversity Outlook 3, http://www.cbd.int/gbo3/.

12. Millennium Ecosystem Assessment http://www.maweb.org/.

13. Global Biodiversity Outlook 3, http://www.cbd.int/doc/publications/gbo/gbo3-final-en.pdf.

14. Mark Strauss, "Looking Back on Limits to Growth," *Smithsonian Magazine*, April 2012, http://www.smithsonianmag.com/science-nature/Looking-Back-on-the-Limits-of-Growth.html#ixzz1sFgJDtdz

15. Oxfam, www.oxfam.org/en/pressroom/2009-04-21/increase-numbers-people-affected-climate-disasters

16. Ian Angus and Simon Butler, *Too Many People*, (Chicago, IL: Haymarket Books, 2011).

17. Les Roopanarine, "Sahel Hunger Crisis Worsens as Aid Agencies Admit Massive Cash Shortfall," *The Guardian*, April 23, 2012, http://www.guardian.co.uk/global-development/2012/apr/23/sahel-food-crisis-aid-cash-shortfall.

18. Peter N. Spotts, "Little Time to Avoid Big Thaw, Scientists Warn," *Christian Science Monitor*, March 24, 2006, http://www.csmonitor.com/2006/0324/p01s03-sten.html.

19. Gary Pfeiffer, Dupont CFO, 340% increase in share value paralleling 60% reduction in environmental footprint, personal communication at speech he gave at a Conference Board conference, 2005.

20. Personal communication, Andrew Winston, Eco-Strategies, http://www.andrewwinston.com.

21. "Walmart Announces Significant Progress toward Ambitious Sustainability Goals in 2012," Global Responsibility Report, April 16, 2012, http://news.walmart.com/news-archive/investors/walmart-announces-significant-progress-toward-ambitious-sustainability-goals-in-2012.

22. Wal-Mart Announces Partnership with Carbon Disclosure Project to Measure Energy Used to Create Products, Press release, September 24, 2007, http://walmartstores.com/pressroom/news/6739.aspx.

23. *Global Responsibility Report*, Walmart, http://www.walmartstores.com/sustainability/7951.aspx.
24. Zogby International poll conducted in November 2005, IBOPE Intelligencia, "New National Poll Finds Americans hold Diverse, Strong, & Increasingly Negative Opinions about Wal-mart." http://www.ibopezogby.com/news/2005/12/01/new-national-zogby-poll-finds-americans-hold-diverse-strong-increasingly-negative-opinions-about-wal/.
25. Walmart data as reported by the American Prospect in May 2011. Rowe, Jonathan. "*The Greening of Wal-Mart*," April 19, 2011, http://www.prospect.org/article/greening-wal-mart
26. This concept was pioneered by Gifford Pinchot in his landmark book. Pinchot, Gifford, *Intrapreneuring* (New York: HarperCollins, 1986).
27. See also Pinchot Gifford, and Pellman, Ron, *Intrapreneuring in Action: A Handbook for Business Innovation* (San Francisco: Berrett-Koehler Publishers, 2000).
28. Marc Gunther, "The Green Machine," *Fortune*, July 27, 2006, http://money.cnn.com/magazines/fortune/fortune_archive/2006/08/07/8382593/index.htm.
29. "GE to Invest $10B More in Ecomagination R&D by 2015," *GreenBiz*, June 24, 2010, http://www.greenbiz.com/news/2010/06/24/ge-invest-10b-more-ecomagination-rd-2015?utm_source=GreenBuzz&utm_campaign=1f92881ad1-GreenBuzz-2010-06-14&utm_medium=email.
30. www.ge.com/ar/
31. Ram Nidumolu, C. K. Prahalad, and M. R. Rangaswami, "Why Sustainability Is Now the Key Driver of Innovation," *Harvard Business Review*, September 2009, Reprint R0909E.
32. "Chief Executives Believe Overwhelmingly That Sustainability Has Become Critical to Their Success, and Could be Fully Embedded into Core Business within Ten Years," *Accenture*, June 22, 2010, http://newsroom.accenture.com/article_display.cfm?article_id=5018.
33. Natural Capitalism Solutions, http://www.natcapsolutions.org.
34. Personal communication, Jeff Hohensee, NCS CEO, December 2008.
35. 1E PC Energy Report: 1E Study, 2009 http://www.1e.com/energycampaign/index.aspx
36. Ariel Schwartz, "Ford Saves One Million Dollars … by Shutting Off Computers," Fast Company, March 23, 2010, http://www.fastcompany.com/1595298/ford-saves-one-million-dollarsby-shutting-off-computers?partner=homepage_newsletter, and Julie Morrison, "Study Shows Turning Off Computer at Night Saves Energy, Big Bucks," *Flint Journal*, April 13, 2009, http://www.mlive.com/news/flint/index.ssf/2009/04/study_shows_turning_off_comput.html.
37. Personal communication, Jeff Hohensee, NCS CEO, December 2008.
38. Verlyn Kinlkenborg, "Light Pollution," *National Geographic*, November 2008, 30 to 60% of energy consumed in lighting is unneeded or gratuitous. See according to Lumina Technologies, Santa Rosa, CA, Survey of 156 California commercial buildings' energy use, August 1996. See also,

"Why in the World Do Businesses Leave Their Lights on at Night?" The Good Human, July 26, 2007, http://www.thegoodhuman.com/2007/07/26/why-in-the-world-do-businesses-leave-their-lights-on-at-night/.

39. Andrew Revkin, "Greens Debating Tactics Instead of Ideas," *New York Times*, August 5, 2010, http://dotearth.blogs.nytimes.com/2010/08/05/greens-debating-tactics-instead-of-ideas/.

40. "Gearing for Growth," Economist Intelligence Unit, 2011.

41. Gallup, http://www.gallup.com/consulting/52/employee-engagement.aspx.

42. Gallup-Healthways Well-being Index, www.greenbiz.com/blog/2012/08/30/fourstages-employee-engagement-sustainability, quotes in http://www.well-beingindex.com/.

43. Teresa Amabile and Steven Kramer, "Do Happier People Work Harder?" *New York Times*, September 3, 2011, http://www.nytimes.com/2011/09/04/opinion/sunday/do-happier-people-work-harder.html?_r=3.

44. 2007 MonsterTrak survey of recent graduates, see "Working for the Earth: Green Companies and Green Jobs Attract Employees," *GreenBiz*, October 16, 2007, http://www.greenbiz.com/news/2007/10/16/working-earth-green-companies-and-green-jobs-attract-employees.

45. Shelley DuBois, "How Going Green Can Be a Boon to Corporate Recruiters," *Fortune*, June 2, 2011, http://tech.fortune.cnn.com/2011/06/02/how-going-green-can-be-a-boon-to-corporate-recruiters/.

46. Eric Omer, "It Costs How Much to Replace an Employee?" Ezine Articles, http://ezinearticles.com/?It-Costs-How-Much-to-Replace-an-Employee?&id=2555834.

47. "Environmental Indicators, Modelling and Outlooks: OECD Environmental Outlook to 2050: The Consequences of Inaction," OECD, http://www.oecd.org/document/11/0,3746,en_2649_37465_49036555_1_1_1_37465,00.html.

48. Fiona Harvey and Damian Carrington, "Governments Failing to Avert Catastrophic Climate Change, IEA Warns," *The Guardian*, April 24, 2012, http://www.guardian.co.uk/environment/2012/apr/24/governments-catastrophic-climate-change-iea?newsfeed=true.

49. Chrystia Freeland, "What's BP's Social Responsibility?" *Washington Post*, July 18, 2010, http://www.washingtonpost.com/wp-dyn/content/article/2010/07/16/AR2010071604070.html?hpid=opinionsbox1%20.

50. "Ernst and Young Business Risk Report 2010: Top Ten Risks for Business," Ernst and Young, http://www.ey.com/GL/en/Services/Advisory/Business-Risk-Report-2010—Business-risks-across-sectors; "'Radical Greening' Seen as Top 10 Business Risk," Environmental Leader, August 4, 2010, http://www.environmentalleader.com/2010/08/04/radical-greening-seen-as-top-10-business-risk/.

51. Blue Earth Network, http://www.blueearthnetwork.com/.

52. World Business Council for Sustainable Development WBC SD, http://www.wbcsd.org/home.aspx.

53. This term was created by Walter Stahel of the Product Life Institute, http://www.product-life.org/en/cradle-to-cradle.

54. For more information on this, see the Biomimicry Institute, http://www.bio-mimicryinstitute.org/.

55. This term was created by Walter Stahel: http://www.product-life.org/en/node. For more information on China's adoption of Stahel's concept, see Indigo Development, http://www.indigodev.com/Circular1.html.

56. Janine M. Benyus, "Biomimicry," 1997, http://www.biomimicryinstitute.org/.

57. Calera, http://calera.com/index.php

58. Nonny de la Pena, "Sifting the Garbage for a Green Polymer," *New York Times*, June 19, 2007, http://www.nytimes.com/2007/06/19/science/19poly.html?_r=1&oref=slogin

59. Personal communication, Ray Anderson, Wingspread Conference, June 2008.

60. The master of this is Dr. Pavan Sukhdev, whose TEEB project, The Economics of Ecosystems and Biodiversity, is the leader in creating this new approach.

61. At a minimum, a wise civilization ensures that its activities leave a safety margin to keep all critical ecological services undamaged. This approach, called the Precautionary Principle, states that if the risk from a particular activity is sufficiently high, a proponent will be required to prove that it is safe before being allowed to go forward. The Principle is now being written into European Union legislation (http://ec.europa.eu/dgs/health_consumer/library/pub/pub07_en.pdf). A few American cities have followed suit, with San Francisco formalizing the Precautionary Principle into its governance. In general, however, American business has persuaded legislatures that an activity should be allowed unless it is proven to be harmful (http://ec.europa.eu/environment/docum/20001_en.htm).

62. BGI, http://www.bgi.edu; http://www.aashe.org/resources/sustainability-busi-nessmanagement-programs; Education Revolution, http://www.educationrevo-lution.org/.

63. First introduced by L. Hunter Lovins of Natural Capitalism Solutions in 1997, this approach contrasts with John Elkinton's Triple Bottom Line, in which companies tend to bolt programs to protect people and planet on to business as usual, becoming cost centers. The IBL argues that integrating more sustainable practices throughout the company is the driver of greater profitability.

64. ICAEW, http://www.icaew.com/en/technical/financial-reporting/other-reporting-issues/narrative-reporting/iirc-publishes-initial-proposals-for-inte-grated-reporting-framework. Hunter Lovins spoke on the IBL concept at a symposium at ICAEW headquarters in London, November 2009.

65. Robert Eccles and Michael Krusz, *One Report: Integrated Reporting for a Sustainable Strategy* (New York: Wiley, 2010).

66. Personal Communication: Hugo Steensma, former senior vice president, Sustainable Asset Management, and Sasha Millstone, senior vice president of the Millstone Evans Group of Raymond James and Associates, Inc. Boulder Colorado. See also GS Sustain, June 22, 2007, http://www2.goldmansachs.com/ideas/environment-and-energy/gs-sustain/index.html, which found that ESG indicators are "a good overall proxy for the management quality of companies relative to their peers."

67. "Paul Krugman: King of Pain," Economist's View, September 18, 2006, http://economistsview.typepad.com/economistsview/2006/09/paul_krugman_ki.html.
68. "Nic Marks: The Happy Planet Index," dotSUB, http://dotsub.com/view/e0269cf5-d62f-40a5-abdf-704e5b0c4f1e.
69. Robert Kennedy, remarks at the University of Kansas, March 18, 1968, John F. Kennedy Presidential Library and Museum, http://www.jfklibrary.org/Historical+Resources/Archives/Reference+Desk/Speeches/RFK/RFKSpeech68Mar18UKansas.htm.
70. J. Stiglitz, A. Sen, and J.-P. Fitoussi, *Report by the Commission on the Measurement of Economic Performance and Social Progress*, 2009, http://www.stiglitz-sen-fitoussi.fr/documents/rapport_anglais.pdf.
71. Jon Gertner, "The Rise and Fall of the GDP," *New York Times*, May 16, 2010, http://www.nytimes.com/2010/05/16/magazine/16GDP-t.html?sq=gdp&st=cse&scp=2&pagewanted=print.
72. "Nic Marks: The Happy Planet Index," TED: Ideas Worth Spreading, July 2010, http://www.ted.com/talks/nic_marks_the_happy_planet_index.html.
73. The New Economics Foundation, http://www.neweconomics.org/.
74. Nic Marks: The Happy Planet Index," TED.
75. Gross National Happiness, http://www. grossnationalhappiness.com.
76. E. F. Schumacher, *Small Is Beautiful* (New York: Harper Perennial, 1959, 1973, 1989), http://www.ecobooks.com/books/smbeaut.htm.
77. Don Duncan, "Economists Appraise Bhutan's Happiness Model," *San Francisco Chronicle*, December 4, 2008, http://www.sfgate.com/cgi-bin/article.cgi?f=/c/a/2008/12/04/MN3C14CTN5.DTL.
78. The State of the USA, http://www.stateoftheusa.org/.
79. Duncan, "Economists Appraise Bhutan's Happiness Model."
80. Sustainable Measures, http://www.sustainablemeasures.com/.
81. Duncan, "Economists Appraise Bhutan's Happiness Model.,"
82. Personal communication, Bhutanese Minister of Energy, Manila, the Philippines, September 9, 2009.
83. Well-Being and Happiness: Defining a New Economic Paradigm, http://www.2apr.gov.bt/#; Hunter Lovins chaired the Civil Society Working Group at this meeting.
84. *Towards a Circular Economy*, Ellen MacArthur Foundation, http://www.ellenmacarthurfoundation.org/about/circular-economy/towards-the-circular-economy.
85. Transition Network, http://www.transitionnetwork.org/.
86. Martin Burund, "Living a Green Dream on Danish Island," Planet Ark, August 19, 2008, http://www.planetark.org/dailynewsstory.cfm/newsid/49847/story.htm.
87. "German Village Achieves Energy Independence … And Then Some," BioCycle, August 2011, http://www.jgpress.com/archives/_free/002409.html.
88. "International Community Power Conference Set for 3–5 July in Bonn, Germany" Wind-Works, January 5, 2012, http://www.wind-works.org/coop-wind/CitizenPowerConferencetobeheldinHistoricChamber.html.

89. J. Matthew Roney, "Solar Cell Production Climbs to Another Record in 2009, Earth Policy Institute," Renewable Energy World, September 24, 2010, http://www.renewableenergyworld.com/rea/news/article/2010/09/solar-cell-production-climbs-to-another-record-in-2009.

90. "Paying for Renewable Energy: TLC at the Right Price —Achieving Scale through Efficient Policy Design," Deutsche Bank, December 2009, http://www.dbcca.com/dbcca/EN/investment-research/investment_research_2144.jsp.

91. Interview with Bill Clinton, Part 2, *The Daily Show with Jon Stewart,* November 8, 2011, http://www.thedailyshow.com/watch/tue-november-8-2011/bill-clinton-pt--2.

92. In May 2010, Hunter spent an hour with Achim, advising him on how such a summit might be configured. "'Global Green New Deal'—Environmentally Focused Investment Historic Opportunity for 21st Century Prosperity and Job Generation," United Nations Environmental Programme, http://www.unep.org/documents.multilingual/default.asp?documentid=548&articleid=5957&l=en.

93. Joseph E. Stiglitz, "Reversal of Fortune," *Vanity Fair*, November 2008, http://www.vanityfair.com/politics/features/2008/11/stiglitz200811, see also Joseph E. Stiglitz, *Freefall: America, Free Markets, and the Sinking of the World Economy* (New York: W. W. Norton & Company, 2009).

94. Peter Hilderson, Jones Lang LaSalle, Asia Pacific, Energy and Sustainability Services *Global Sustainability Perspective*, Feb 2012.

95. Lester Brown, China Forcing World To Rethink Its Economic Future, Earth Policy Institute, 5 Jan 2006 http://www.earthpolicy.org/Books/PB2/index.htm

Chapter 3

Lean and Green: Principles and Strategies

What Is Lean and Why Is It Green?

The term *lean* was first coined by John Krafcik—now president and CEO of Hyundai Motor America—in 1988. Krafcik, who was researching the comparative performance of automotive assembly in Japan versus the West, used the term to describe Toyota's ability to do much more with much less[1] in his master's degree dissertation at the Massachusetts Institute of Technology. The term *lean* was later popularized by Womack and Jones, especially in their seminal book *The Machine that Changed the World*.[2] That machine was, of course, the Toyota Production System. The benchmark studies presented in this bestseller highlighted a substantial gap between the Japanese automakers—especially Toyota—and the rest of the world, in terms of both quality and productivity. For example, they observed that the number of hours it took to build a Lexus was less than the time spent to rework a top-of-the-line luxury German car at the end of the production line after it was manufactured.[3] Rather counterintuitively, they also found a strong correlation between higher productivity and better quality, meaning that achieving best-in-class quality does not require extra effort such as rework at the end of the production line. Far from it, in fact, carmakers with the highest productivity were also best in quality. At last, quality was proven to be free, exactly as the quality gurus of 1960s and 1970s preached.

Not finding American executives interested in their ideas, gurus such as Dr. W. Edward Deming went to Japan to help rebuild industries after World

War II. Unlike in America, their ideas were widely adopted by Japanese executives, above all Toyota, resulting in the enormous gap in performance. Toyota had created a production system that went on to become a legend in modern management and Toyota itself to become an icon and arguably the most researched company in the world: a flexible production system that made exactly the cars customers wanted, in a fraction of the time, at much lower cost, and with far better quality, while enriching employees working lives and learning to minimize its negative impacts on the society.

But Toyota never called their production system *lean*, a word that can be one of the most unfortunate in the management lexicon. *lean* was originally used to describe Toyota's ability to do more with less and what seemed to be a relentless focus on waste elimination. Nonetheless, as the lean community learns more about the Toyota Production System (TPS), it has become increasingly obvious that TPS is all about continuous *value enhancement* rather than perpetual *waste reduction*. In the Toyota Production System, waste is crucially defined in relation to value. *Waste elimination* is a virtue as long as it leads to value enhancement, such as inventory reductions, leading to shorter delivery times for customers at a given or even lower cost. *Lean* carries a strong efficiency connotation pointing to reduction of fat; the Toyota Production System is all about effectiveness, which means building muscle. Efficiency is doing things right, but effectiveness is doing the right things. Toyota learned to do the right things first and then *to do the right things righter and righter*—not just in an economic sense, but also for the society and environment. So the dubbing of the Toyota system as lean was simply a Western manipulation influenced by the mass production perceptions.

In the words of Jim Womack,[4] "from [the] society's standpoint the reason organizations exist is to create value for consumers. So successful organizations … are the ones that actually solve consumer problems … . People are looking in the wrong end of the telescope when they think that the purpose of the organization is cost reduction. So much of the lean movement up to this point has been focused, I think in an unbalanced way, on cost reduction as opposed to value maximization which is what the customer really wants."

In this book we use the term *lean* to refer to the ethos of continuous improvement that underpins TPS rather than mere waste reduction. Indeed, when lean is translated to other languages it often takes on different meanings closer to its origins. In French the literal translation of lean results in *gestion maigre*, meaning "skinny or anorexic management." It is interesting that this translation is sometimes used in the French-speaking parts of Canada, maybe showing mass production influences. However, the common phrase for lean management in French is *gestion au plus juste*, meaning "the most appropriate or justified management." While Germans and Italians prefer to use the English phrase, Persians translated *lean production* as "pure or limpid production," which again provides a value connotation as well as emphasizing the simplicity aspects of TPS. But even more interesting, in Mandarin, *lean* is translated as *Jing Yi*, which means the "*essential core for the benefit of all people.*" For the Chinese, *lean* goes beyond value creation for the customer to include a wider range of stakeholders even beyond the immediate community to include "very many people" or in plain English, the entire human society. It gets even better! Jing Yi is also the first half of the idiom *Jing Yi Qiu Jing*, which means "perfect, but aiming for even better"; in other words, the Mandarin translation also means *relentless pursuit of perfection*.

The Chinese translation of lean simply takes us back to the original intentions of the fathers of the Toyota Production System and to the teachings of Dr. Deming. Value enhancement (*Jing* or the essential core) for consumers is to receive cars that are better priced and more reliable; value for the employer (Toyota) is in longevity and profitability; value for the employees is that they are not robbed of their pride of workmanship; and value for the wider society is clean air and access to natural resources for future generations. Is there a better way to describe sustainability?

But what is the Toyota Production System? The system that we, rightly or wrongly, regard as lean.

Toyota, as well as many other Japanese industries, learned from Dr. Deming to focus on doing the right things rather than *doing the wrong things righter*. Deming was a systems thinker. He spoke about the 94/6 rule, where 94% of problems can be traced to the process and only 6% to the person. But it is the person that is often measured and even blamed for problems. Deming said that any meaningful improvement comes from action on the system, the responsibility of the management, and that wishing, pleading, begging, or even threatening employees to do better is entirely futile until we have fixed the system.[5] It is the system of work that we need to act upon to improve, not the output and certainly not blaming the people.

One of Deming's fascinations was the importance of "real knowledge" or *understanding of work*. He emphasized the need to be close to the workplace and also to measure variation embedded in any operation. What Deming meant by gaining knowledge was for managers to get out of their offices and become intimately familiar with the system of work—something so rare in command-and-control organizations of today.[6] Deming explained the importance of enabling employees to improve their own operation and to instill the organization with the pride of workmanship. These points were, of course, quite the opposite of the command-and-control paradigm of the mass production era whereupon the worker is a powerless adjunct of the machine expected to do whatever is asked without thinking or understanding, let alone being able to act on the system to improve it. But above all, Deming emphasized the importance of changing our thinking before applying the tools and techniques for improvement.

Toyota put all of this to work on a grand scale.

Not so surprisingly, when Western executives first came across Toyota's remarkable performance gap, they looked at the tools and methods Toyota was deploying with little regard for the fundamentally different ways of thinking (paradigm) that underpinned those techniques. lean, therefore, soon became a toolkit. Toyota was surprisingly open in sharing its approach, inviting Western executives inside their factories and even going into joint ventures with established Western manufacturers in North America. But the old school *efficiency thinking* lens, through which the Western managers looked at the Toyota Production System, witnessed the difference in output, but failed to see the whole system thinking therein. Toyota simply thought about manufacturing in a different way; it was the *systems thinking* that mattered—not the toolkit. Einstein famously said we cannot solve problems by using the same kind of thinking we used when we created them.

So Toyota's superior economic performance is linked to their novel way of thinking, which we regarded as systems thinking. It must be said that Toyota's thinking, in its roots, is also in tune with the needs of the society and environmental considerations. An obvious example is in the *Taguchi* approach to quality management. Toyota adopted the thinking of the influential quality guru, Dr. Genichi Taguchi, in the 1950s and 1960s. Directly influenced by Deming's teaching, Taguchi detested the idea of quality linked to postproduction inspection. He saw quality as an integral part the process and emergent from the design stage, leading him to define quality in terms of the *loss imparted by the product to the society from the time the product is shipped from the supplier.*[7] So, good quality means eliminating losses to the society, which can be caused by breakages, repairs, delays, customer dissatisfaction, and any other type of waste linked to a quality issue. Whereas the old school definition of *quality* provides a range of specification limits to define good quality, Taguchi's loss function implies that the company should

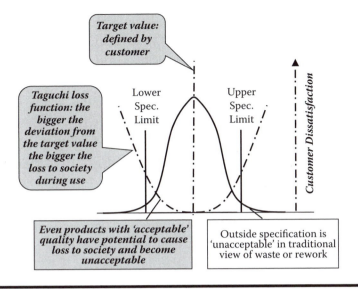

Figure 3.1   Taguchi concept of quality: Loss to the societies.

strive to always meet exactly the nominal value, or in other words, be right in the middle of the specification limits for all products.

Figure 3.1 is a depiction of the Taguchi principle. The traditional definition of quality implies that any product that is within specifications is acceptable; in other words, good quality. Any attempt to make all products to be exactly in the middle of the specification will simply cost extra without any tangible benefits. Or at least, such goes the efficiency logic. However, Taguchi's loss function concept says that loss to the society is proportional to the deviation from the target value, which is in the center of the specification limits and is defined by the customer. The larger the deviation from the center of the specification limits, the higher the chances of loss to the society during its use—for example, product breakages. Taguchi's *effectiveness logic* implies that we should do the right thing and minimize the potential losses during the product life cycle by always meeting the center of the graph.

Interestingly, the Taguchi method of quality not only costs no more, but in fact costs much less from the whole system point of view. Here is a story to explain why. When Ford began purchasing Toyota models for reengineering in the late 1960s, they were amazed to find that when disassembled, they were 100% *snap-fit*—a term in the Ford assembly line to define those connecting parts that could be assembled without the use of a rubber mallet. The assembly plant manager, astonished by the fact that a competitor could achieve such a level of reliability (so far-reaching for Ford at the time) invited Ford senior executives to witness the result for themselves. Having observed the startling difference in quality, they dismissed the benefit as something that the customer would not notice and that would cause extra unnecessary cost. What Ford executives at the time did not realize was that by snap-fitting all assembly parts, Toyota was manufacturing far more reliable cars, which meant more satisfied customers, fewer repairs or lower warranty cost, fewer recalls, less rework, and even less impact on the society. It would take Ford another couple of decades to realize that Toyota was actually reducing the total cost while also growing their market share, eventually becoming one of the most profitable and sustainable organizations on the planet in spite of competing in a very mature sector.[8]

As we can see from this example, Taguchi's definition of loss to the society also includes the environmental and social impacts during use. This has been at the very heart of TPS since its inception in the 1950s. When an unreliable product component breaks down and requires replacement, it creates social loss (inconvenience) and environmental loss (carbon emissions embodied in the replacement component, which would not have been released if the component had been reliable) as well as economic loss (cost of replacement). *Taguchi's counterintuitive logic to encompass social waste, though it may seem altruistic and impractical, turns out to be far more profitable.* True lean, manifested in the Toyota Production System, is concerned with doing the right things from a whole system point of view and enhancing value for all people proximate or distant, in the present or future. In the Toyota sense, lean and green are part of the same approach. In the following chapters we talk more about the latest lean and green achievements of Toyota, especially in a dedicated case study in Chapter 4.

While we have some reservations around the misinterpretations of TPS, sometimes reducing it into a toolkit, we think that the lean movement overall has had a positive impact on changing conventional management thinking. lean and green practitioners should be aware of the "toolheads"[9] who indiscriminately try to force tools and methods that worked well in one environment into any situation. While it may sound controversial, there are no codified shortcuts or a standard toolkit for becoming lean and green. Becoming lean and green is a highly contextual process that requires commitment and hard work. Highly prescriptive solutions are highly likely to "straightjacket" us into failure. In the following, as we set out to offer various tools, techniques, and methods for lean and green, we need to caution against the use of tools without nurturing a true lean and green paradigm.

There simply is no replacement for thinking.

So far we have established that lean is more than just waste elimination, that lean is about adopting a new paradigm—a new way of thinking about quality, productivity, efficiency, and value creation. Toyota Motor Company defines the two pillars of the Toyota Way as *continuous improvement* and *respect for people*.[10] Those lean practitioners who have achieved more than fleeting success with lean will know that to achieve economic success, you require a major focus on your people. We will come back to the meaning of real lean and the Toyota Way when we discuss Toyota's own lean and green achievements in Chapter 4.

But even if we take the most simplistic definition for lean, which is waste elimination, there are obvious overlaps with green. It was just after the turn of the millennium that Jim Womack wrote in an online blog: "Lean thinking must be 'green' because it reduces the amount of energy and wasted by-products required to produce a given product. Indeed, examples are often cited of reducing human effort, space, and scrap by 50 percent or more, per product produced, through applying lean principles in a manufacturing facility. ... This means that lean's role is to be green's critical enabler as the massive waste in our current practices is reduced."[11] For those companies that already have thriving lean programs, taking a further step to encompass the environment seems to be only logical, especially because it can also yield more economic and social benefits.

To think in very simple terms, *lean* might be described as understanding the customer's needs and values, and then reviewing the value streams that produce them so that the eight wastes of lean can be minimized. The eight waste of lean were originally introduced by Taiichi Ohno,[12] the father of Toyota Production System. The original eight lean wastes are (Figure 3.3):

■ Overproduction
■ Defects
■ Unnecessary inventory
■ Unnecessary transport
■ Waiting (idle people and machines)
■ Inappropriate processing
■ Unnecessary motion
■ Wasted human potential

Similarly, *Green* can be described as understanding the society's needs and values and then reviewing the entire system that delivers them so that the environmental wastes can be minimized. So what is the difference? Well,

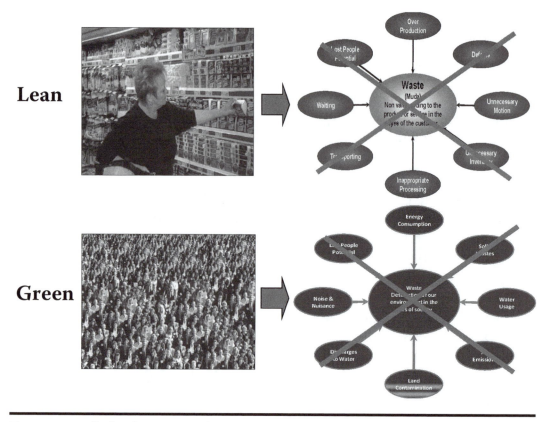

Figure 3.2 Eliminating Lean and Green wastes.

apart from the fact that individual customers are multiplied to become society and the environmental wastes have a slightly different character than the traditional lean wastes, not a lot (see Figure 3.2).

In this book we define the eight wastes of green (Figure 3.4) as follows. We later come back to these eight wastes and develop our lean and green toolkit in the next chapter.

- Excessive energy usage
- Physical waste (solid or liquid)
- Excessive water usage
- Air emissions (most importantly the greenhouse gases)
- Land contamination
- Discharges to water and effluent
- Noise and nuisance
- Lost people potential

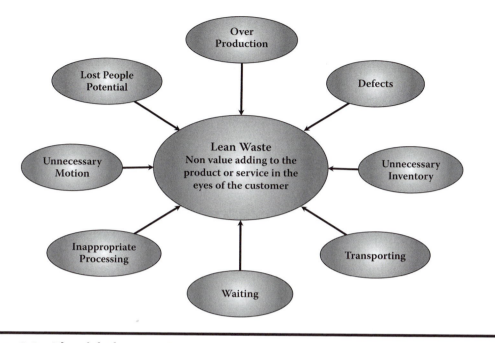

Figure 3.3   The eight lean wastes.

As we saw in the first chapter, progress seems to have been somewhat slow on the lean and green front. This is despite the relentless march of lean thinking across industry sectors and the heightened awareness of green through high-profile activities such as the Stern Review.

A key reason for this lack of progress is the lack of a systematic approach for integrating lean and green into our business system. To help, in this book we offer the Lean and Green Business System model shown in Figure 3.5. Think of the four elements of the model as the pieces of a jigsaw puzzle. The benefit of getting all of the pieces in the right place at the right time is that you are able to see the whole picture. The same is true of the Lean and Green Business System model. Over the course of a transition to develop a lean and green organization, we must systematically complete the whole of the jigsaw puzzle to gain full benefits from our investment.

In order to create a truly lean and green business, we need to excel in all four areas, for example, leaning and greening our business processes, such the production process or the product development process, is crucial (right-hand part of the model). Though, leaning and greening processes alone, without attention to *people engagement* and *strategy*, is hardly sustainable. A lean and green organization also needs to engage people at all levels (bottom part of the model). Moreover, we need to deploy lean and green at a strategic level and to align the day-to-day operations with the company's

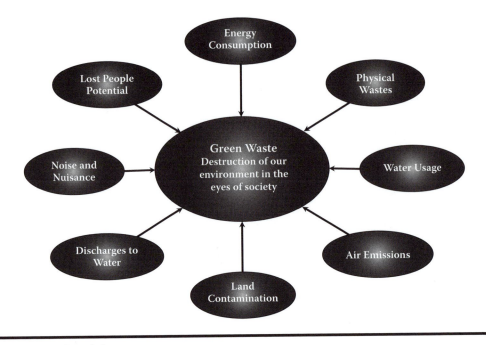

Figure 3.4 The eight green wastes.

lean and green intentions (top part of the model). It is equally important to look beyond the four walls of the factory into the supply chain, both upstream and downstream (left part of the model). Lean and green supply chain management is especially important as the social and environmental impacts of products and services often occur elsewhere in the end-to-end product life cycle.

The Lean and Green Business System model gives a systematic way for organizations to design their journey to eliminate green wastes and to generate green value. Currently, many organizations seem to be void of systematic methods for continuous green improvement. Today, too many environmental management systems are reduced to "postmortem" audits. This is very reactive. Green interventions are skewed in the direction of one-off fixes and end-of-pipe solutions involving few people. Lean's greatest contribution to environmental management can be in creating engagement and alignment through bottom-up continuous improvement and action learning.

The lean community is increasingly using systematic improvements while also paying attention to cultural and leadership issues. Similarly, the lean community is widely familiar with orderly improvement methods such as value stream mapping or road mapping techniques. *The Lean and Green Business System model provides the basis for organizations to design their systemic roadmap for becoming lean and green by understanding their current*

Figure 3.5   Lean and Green Business System model for creating a lean and green enterprise.

state against each of the four areas and then targeting a future state for improvement. In the next section, we provide methodological solutions for mapping processes, creating current state and future state maps, and moving to implement the targeted future state.

The next section of this book consists of four chapters. Chapters 4 through 7 each take on one element of the Lean and Green Business System model. As illustrated in Figure 3.5, we will draw upon a number of case studies to explain how leading organizations are integrating lean and green into their daily operations. But before moving on to explain each element of the model, we briefly discuss the concept of sustainability and explain why it is so fundamentally important to our modern economy. We then take a look at the evolution of lean and green, the attitudes toward them, and the current levels of adoption.

Sustainability and the Case for Lean and Green

The case for lean and green is twofold. First and foremost, going green makes perfect business sense. Lean and green are innately connected. By radically increasing resource efficiency of the firm and by creating a culture where the entire company, its supply chain, and even its consumers strive toward less resource intensity, fewer wasteful activities, and more value-adding activities, you *will* discover new horizons in competitiveness. In Chapter 2, we explored the business and economic case for lean and green.

There is also an ethical case for lean and green. Fortunately, the ethical case makes perfect business sense too. The ethical case for lean and green is often identified with the concept of sustainability, which is about convergence of the economic performance of the firm with social and environmental considerations in a state of harmony. Here we take a quick look at this ethical case, briefly discuss the background to the concept of sustainability, and highlight its principles, followed by an overview of some of the most influential schools of thought, which suggest practical solutions and call for progress toward a more sustainable economy.

Many social and environmental challenges have accrued since the industrial revolution, severely constraining the pure economic development of the human race. Much progress on many economic fronts is compromised by entrenched corruption and environmental degradation. As a result, there is a profound concern among both industrialists and policy makers regarding the world's natural capacity to cope with the existing economic rates of growth. The influential World Bank economist Herman Daly[13] breaks this down, explaining that we have one planet inside which there is a finite amount of resources and that the human economy has to exist within these physical limitations. Currently, we consider our economy as a separate entity from the natural ecosystem that supports it; this rather bold disassociation of the economy and ecology is what the renowned ecologist, David Ehrenfeld, refers to as the "arrogance of humanism."[14] Such a short-sighted *dualistic* economic view, which we have inherited from the time of enlightenment, externalizes the costs imposed by the industry on both the environment and the noncontemporary and nonproximate human beings. The paucity of attention to the fact that any economic organization is embedded in the context of the natural environment and the injudicious externalization of the economic services provided by the ecosystem have had dire consequences for humans, with possibly much worse to come in the future.

The human economy works by taking energy and resources and turning them into goods and services supposedly contributing to humans' welfare (happiness). This process creates waste, some of which can be recycled to replenish the finite supply of resources, but most of which accumulates in nature as pollutants.

Notably in our economic system, the monetary value of goods and services are set by the balance of supply and demand in markets. Yet prices set in markets traditionally omit certain costs and benefits that do not have a market price—the so-called externalities. Externalities often come from stakeholders lacking power to make their value or cost recognized, for example, detrimental impacts of a large corporation on its employees' health or the natural resources used and the pollutants produced. As a consequence of externalizing the environmental costs, our industrial practices are profoundly disturbing the ecological balance of the planet in many different ways at the same time that disparities are widening in human society. Some of these challenges are climate change, eutrophication, ozone layer depletion, loss of biodiversity and deforestation, extreme poverty, malnutrition and undernourishment, and acute social inequality, to name but a few. Full review of these challenges is beyond our scope; the point is that the human economic system needs to *internalize* such externalities.

In order to do so, it is necessary to begin to measure and understand the environmental costs and benefits, to introduce "currencies" that capture the environmental values alongside monetary currencies. For example, the carbon footprint is one such currency. There is a general consensus among environmentalists that climate change is one of the most pressing environmental crises of all. The United Nation's Intergovernmental Panel on Climate Change (IPCC) defines climate change as "any change in climate over time whether due to natural variability or as a result of human activity."[15] There is a strong consensus among scientists that most of the global warming observed over the last 50 years is attributable to the increase in the concentration of greenhouse gas emissions due to human economic activities.[16] According to the most recent IPCC report, warming of the climate system is unequivocal and it is more than 90% likely to be due to the increase in anthropogenic greenhouse gas concentrations. Scientists believe that unless urgent and strenuous mitigations are put in place now, it is almost certain that by the end of the century, global temperatures will rise by between 1.1°C and 6.4°C above the current levels,[17] which can have dire consequences such as rising sea levels, changing weather patterns, extreme weather, droughts, floods, loss of wildlife, mass migrations, and food shortages. Global warming and other ecological crises are also coupled with the accelerating global population growth rate, meaning that more people will be competing over fewer natural resources.

Concerns regarding the planet's capacity to deal with the global economic growth were initially heightened during the 1960s and 1970s. The publication of Rachel Carson's *Silent Spring* in 1962 and the Club of Rome's *Limits to Growth* analysis in 1972[18] drew a striking picture of the omens of the environmental doom and raised public attention to the issues surrounding corporate social responsibility. Eventually, this trend led to the adoption of a resolution in 1983 by the United Nations' General Assembly to establish a commission for seeking ways to move forward. Hence, the World Commission on Environment and Development (WCED) was established and chaired by the former Norwegian Prime Minister Gro-Harlem Brundtland. WCED in its concluding report, *Our Common Future*[19] introduced the concept of sustainable development as "development which meets the needs of the present without compromising the ability of future generations to meet their own needs."

The commission's definition of sustainability implies *a state of harmony* aiming at the preservation of the vital assets of the future generations, such as a sound ecosystem and balanced human society. It means that economic expansion should not be at the cost of eroding our medium- or long-term capital. It simply tells us that we should learn to live off our income rather than liquidating our environmental and social capital and consuming it at the expense of future generations.

At the same time, it provided a more balanced view of how businesses can relate to the environmental concerns, and it was a shift from the business-bashing green movements of the past toward a more industry-friendly environmental stance. A bit like creating lean in our company, adopting the concept of sustainability requires a paradigm shift or maybe even a leap of faith. Luckily for us, there is ample evidence that the new paradigm makes even more sense economically. Also, we know that if we don't adopt more sustainable practices, we are far more likely to fail on both a micro and macro scale.

A seminal report published in 2006 demonstrates that the climate change, resulting from human activities, could shrink the global economy between 5% and 20% now and forever.[20]

A very good way of conceptualizing sustainability is the three pillars model, also referred to as the triple bottom-line model or *people-planet-profit*.[21] The overlap among the three pillars in the middle of the diagram (Figure 3.6) relates to the notion of sustainability. The model stresses the long-term compatibility of the economic, social, and

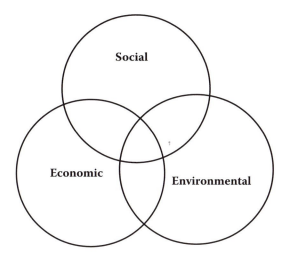

Figure 3.6 The three pillar model for sustainability.

environmental aspects of human activities while acknowledging potential trade-offs in the short term.

But, unfortunately, the concept of sustainability soon became something that everyone liked but nobody knew how to implement. While it was sufficiently vague to allow for political and industrial consensus, it did not provide any practical guidance on what to do or how to implement it. It has therefore been criticized ever since its conception for being, arguably, too broad and therefore impractical.

But isn't that just like lean?

By the same token, the concept of lean—doing more with less or making tomorrow better than today—is too broad and has indeed been criticized for being too abstract and unusable. Both concepts are open to interpretations and both phrases are ambiguous ones that can be inferred in many different ways. For example, lean thinkers are often at pains to explain that lean is not simply about cutting waste or reducing fat, but that the emphasis needs to be on creating value or building muscle.

Both *lean* and *sustainability* can be viewed as some sort of "holy grail" state to which one can aspire. It is interesting that both terms were coined in the 1980s. Nonetheless, one cannot help noticing that the concept of lean, in spite of much skepticism in the early days and criticism for its supposed abstractness, has enjoyed much wider adoption on its own and with far less political wind behind it. This is why we think lean practitioners have something to share with the green community on how to successfully relate with business realities and how to show practical relevance. Of course, there have been companies that have ignored the concepts of lean and continuous improvement. They have paid the price, doubtless. The same goes for green, except that this time we will all be paying the price, not just the companies who ignore its principles.

Clearly, a critical prerequisite for wider adoption of sustainability is practical guidance backed up by real-life examples from companies that have benefited from putting sustainability into practice and have managed to create the favorable conditions portrayed by its vision, that is, simultaneous economic, environmental, and social prosperity. One finds solace in knowing that many companies around the world have successfully begun their green journey and implemented significant change. So there are many examples and we will discuss a handful in the following text.

As for lean, companies have been engaging the workforce, drawing on new ways of thinking, educating at all levels, moving away from tired paradigms of the command-and-control era, and striving for perfection. There has been considerable help from the promoting community including lean thinkers, visionaries, publishers, consultants, and nonprofits. This has been a key enabler for the lean movement. At a practical level, the adoption of sustainability concepts has not been nearly as widely popular or mature as lean. However, the promoting community and the visionary thinkers are equally contributing to the adoption of newer and better ways of working. We will take a closer look in the following sections.

A major obstacle to the integration of the sustainability concept into corporate strategies is that, while there have been numerous visionary strategies and conceptions of what sustainability might look like, there have been fewer practical examples put forward for integrating the concept of sustainability into businesses' day-to-day practices. There remains a great disconnection between the vision of a sustainable industry and its realization. We recognize that operationalization and measurement of sustainability remain in their infancy. Saving the natural environment should become part of the day-to-day operations of companies and their supply chains and not a separate function.

Evolution of Corporate Sustainability Thinking

Sustainability is the capacity to endure. Here, however brief, we review the evolution of sustainability thinking and look at some of the most influential schools of thought. A number of eminent thinkers have discussed the severe social and environmental constraints that have emanated from humankind's colossal economic expansions since the industrial revolution. Corporate social responsibility (CSR) paradigms have evolved over time, from the early concerns regarding the influences of industry on the natural environment (in the wake of Rachel Carson's *Silent Spring* in 1962) to the conception of sustainable development by the UN's World Commission on Environment and Development and its wide acceptance by policy makers and practitioners around the world.

Although the idea of sustainability *per se* is well received, its true essence is rarely implemented. While significant opportunities for win–win solutions exist, external pressures and the need for legitimacy are still the biggest influences for adoption of sustainability. The point of lean and green is to create a more proactive approach to sustainability to seek those win–win opportunities rather than reactive fixes that are driven by compliance or mitigation.

Figure 3.7 traces the evolution of the sustainability agenda from its modern roots in the 1960s through to the present postcrunch world capturing the overarching characteristics of the corporate sustainability theories in the decades since the 1960s.

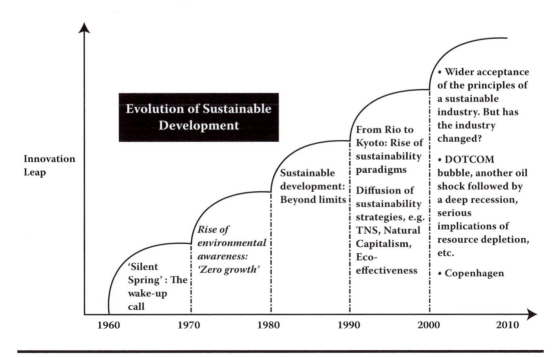

Figure 3.7   Evolution of corporate sustainability paradigms.

During the 1990s, many business solutions emerged mostly aiming to marry the goals of industry with the considerations of sustainable development, for example, *Natural Capitalism*,[22] *Factor Four*,[23] *Eco-efficiency*,[24] *Eco-effectiveness*,[25] and *The Natural Step* in the early 2000s.[26] *The Kyoto Protocol* was adopted in 1997 in Kyoto, Japan, and has since been ratified by around 190 countries. The significance of the Kyoto agreement is that it was the first real practical step in battling the climatic change, upon which developed countries committed to reduce their greenhouse gas (GHG) emissions by around 5.2% below their 1990 levels by 2010, or to otherwise engage in emissions trading as an alternative mechanism.

It was evident from the UN secretary-general, Kofi Annan's opening address to the World Summit on Sustainable Development (WSSD) in Johannesburg in 2002, that the anti-industry tone of the environmental movements of the past decades had evolved into a much more realistic stance seeking to engage businesses in the greening of the industry. "Today, there is growing recognition that lasting and effective answers can only be found if business joins in partnership and working together with other actors including government and civil society and of course trade unions and we all have to remain fully engaged. We now understand that both business and society stand to benefit from working together."[27]

The dominant discourse in this decade is that business can be good for sustainable development provided that appropriate management practices are adopted. At the same time, following the publication of the fourth IPCC report in 2007 and the Stern Review in 2006 an even stronger consensus is formed that severe global warming is attributable to an increase in the human-induced greenhouse gas concentrations. The Stern Review forecasts that the disruptions caused by climate change can shrink the global economy by 20% by 2035 unless 1% (later raised to 2% to reflect continuing inaction) of the global gross domestic product (GDP) is invested annually in measures to mitigate or respond to climate change.[28] By mid-2008 resource depletion was once more challenging the prospects of global economic development with oil prices soaring to $150 per barrel. Many analysts predicted a deep global recession, partly due to high oil prices, and not surprisingly the global economy plunged into a depression, a state of melancholy from which we are yet to recover and one which has perversely decelerated the sustainability agenda.

By December 2009, the Copenhagen Accord provided an initial agreement on key elements of the global climate change framework for the future, but no detailed commitments were put in place. Negotiations toward a detailed, legally binding agreement have continued and a review of the accord and its implementation should be completed by 2015. Following the partial success of the Copenhagen Accord, the United Nations Conference on Sustainable Development was held in 2012 in Rio de Janeiro, 20 years after the previous Earth Summit in Rio, to secure a renewed political commitment to sustainability. In spite of criticism from nongovernmental organizations (NGOs), the outcome document of the conference was a reaffirmation of the international commitment to make every effort to accelerate the achievement of the previously agreed upon sustainable development goals.[29] The implications for businesses are significant with billions of dollars being allocated for the development of cleaner technologies and adoption of sustainable industrial practices as well as carbon reduction targets for 2020 being agreed upon by several major industries.

In Appendix A, we provide a more detailed overview of some of the key strategies portrayed by industrial visionaries for generating a new sustainable industrial era. Study of these strategies and relevant case examples is the best way—if not the only way—to gain a practical understanding of what sustainability means and what it looks like.

Integrating Lean and Green

In the words of the thirteenth-century Persian poet, Faryumadi, there are four stages of learning and therefore four types of learners:

> One who knows and knows that he knows … His horse of wisdom will reach the skies.
> One who knows, but doesn't know that he knows … He is fast asleep; you should wake him up!
> One who doesn't know, but knows that he doesn't know … His limping mule will eventually get him home.
> One who doesn't know and doesn't know that he doesn't know … He will be eternally lost in his hopeless oblivion!

Companies follow the same four stages of maturity as they learn how to integrate lean and green. We show this in Figure 3.8. As the relationship between lean and green matures in a company, it learns to move from perceiving conflict between the two concepts to realizing benign, synergistic, and eventually symbiotic relationships. We would like to acknowledge

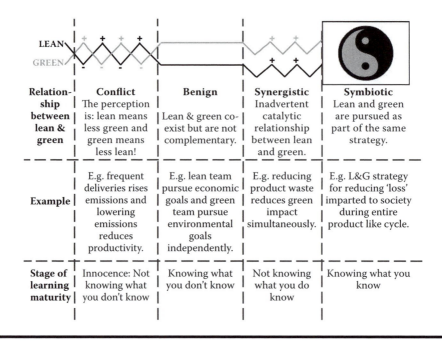

	Conflict	Benign	Synergistic	Symbiotic
Relationship between lean & green	The perception is: lean means less green and green means less lean!	Lean & green co-exist but are not complementary.	Inadvertent catalytic relationship between lean and green.	Lean and green are pursued as part of the same strategy.
Example	E.g. frequent deliveries rises emissions and lowering emissions reduces productivity.	E.g. lean team pursue economic goals and green team pursue environmental goals independently.	E.g. reducing product waste reduces green impact simultaneously.	E.g. L&G strategy for reducing 'loss' imparted to society during entire product like cycle.
Stage of learning maturity	Innocence: Not knowing what you don't know	Knowing what you don't know	Not knowing what you do know	Knowing what you know

Figure 3.8 Stages of lean and green maturity. (Adapted from K. Zokaei et al., *Best Practice Tools and Techniques for Carbon Reduction & Climate Change*, Cardiff: Produced on behalf of CO$_2$ Sense Yorkshire, 2010.)

Martinez, Vasquez, and Zokai who originally contributed conceptualizing an earlier version of this model.

The first stage is when managers see only *conflict* between lean and green; it denotes the start of the journey for most organizations where there seem to be considerable trade-offs between leanness and greenness. Imagine a company that implements just-in-time practices requiring more frequent delivery of goods, which in turn means more miles on the road. This is hardly surprising when one considers centuries of one-dimensional economic growth at the expense of our natural and social capital; it is symptomatic of total innocence about lean and green. In other words, such companies *don't know what they don't know.*

Nevertheless, many companies have already started to find a better balance where applying lean thinking is not necessarily costing the environment. In fact, most companies that we come across can be described as being in this stage of *benign* relationship. These companies separate lean from green and try to avoid any negative effects on the environment caused by their leaner operations, but at the same time, they overlook the potentially catalytic relationship between leanness and greenness. After all, lean and green share the mantra of *tomorrow better than today.* In terms of organizational learning, this can be described as *knowing what they don't know* (or what they don't have). Journeys on a limping mule!

There are many firms across different industries not making the real connection between the continuous improvement team and the environmental function. As soon as companies put aside the silo mentality and begin to see opportunities that lie in connecting green with lean, great amounts of innovative energy are released. Today, this is one of the biggest opportunities missed across industries.

A smaller number of companies, however, leap into an elevated stage of maturity where lean and green are mutually reinforcing. These companies are often fairly mature in either leanness or greenness. However, they have yet to consciously use their lean or green competence to nurture the other one. For example, they might use lean tools and techniques that are aimed at eliminating economic waste, but *inadvertently* benefit the environment. As such, there is a *synergistic* relationship between the two but it is not a systematic lean and green strategy. The awareness of these firms about the synergistic potentials of connecting green and lean is not complete; there is a big prize to be had, but they don't necessarily have a systemic way of pursuing it—a stage of learning maturity that can be described as *not knowing what they do know.*

The final stage of maturity is when a company sees lean and green as part of the same continuous improvement strategy, when lean and green become one and the same thing. These companies adopt a systematic approach to combining and applying lean and green. We call it the *symbiosis* of lean and green. Some of the case studies in the following chapters, cases such as Toyota, Tesco, MAS, Adnams, and Marks and Spencer, are either at this stage or may be somewhere between synergy and symbiosis. These are companies that *know what they know.* Mature firms integrate lean and green at the design stage for both the product and the process, eradicating waste at the root cause rather than dealing with the symptoms. This is exactly what Dr. Taguchi meant when he defined quality in terms of minimizing "the loss imparted to the society from the time the product is shipped," taking a much wider definition than just narrowly focusing on the immediate customer.

In an example, one of us deployed a lean and green toolbox on the shop floor of a multinational manufacturing company to engage lean, green, and engineering teams. In a single weeklong Kaizen blitz event, the improvement team, which consisted of more than 20 engineers and the site general manager (GM), identified 16% of the site's total energy consumption and 15%

of water usage to be saved. Continuous improvement teams were formed around the identified opportunities and an A3 problem-solving approach (a lean technique) was deployed to deliver the results. Financial benefits are in the millions, but more importantly, the likelihood of sustaining environmental benefits are much higher when so many people from different functions and across different levels are actively engaged and taking ownership of the improvement projects. Clearly, this approach needs *not* to be limited to a single firm or a single factory. In a supply chain improvement project, representatives from across the supply chain were put together to map an end-to-end food chain from raw material to the retail shelf. This led to the identification of considerable economic and environmental benefits. To be precise, in excess of £6 million in potential cost savings and similar figures in potential sales uplift as well as thousands of tons of CO_2 and thousands of cubic meters of water savings were identified. In both cases, long-established lean techniques such as value stream mapping were adopted to identify green opportunities and to create more integration between lean and green. We will discuss these approaches in more detail in the following chapters.

Although the lean and green tools and techniques are context specific and the results depend on the problem at hand, one thing is certain: a holistic approach to lean and green will contribute a great deal to both the sustainability and the effectiveness of green improvements, while at the same time providing a more ethical stance for lean as originally portrayed by gurus such as Dr. Deming and Dr. Taguchi.

To sum up, the four maturity stages of conflict, benign, synergy, and symbiosis represent approaches by which a firm can positively (symbiosis and synergy), negatively (conflict), or neutrally (benign) associate economic performance with environmental performance. Of course, it is possible for different divisions within the same company to reach different states of progress at any given time depending on various factors. It is possible for any part of the organization or any individual to go through these stages of learning maturity about the relationship between lean and green.

Over the past two decades, the lean community has attained the wisdom of pursuing operational improvements, regardless of the short-term financial justifications, knowing that a continuous improvement culture pays off over the medium to long term. Companies with mature lean operations know that a detailed cost–benefit case is not required for every step along the journey. On the contrary, they pursue quality and delivery over cost, and customer value and effectiveness over efficiency and *activity cost.* Lean thinkers understand that the long-term benefits of *doing the right things* more than offset the short illusory benefits of *doing the wrong things righter.*

The next step for the lean community is to consciously account for the environment; the initial gains are likely to be more than incremental. At the same time, the green community is maturing in understanding the importance of continuous innovation and systematic methodologies for delivering environmental excellence. At Cardiff University, a research team led by Keivan and Peter carried out a survey into the attitudes toward adoption of lean and green and the associated levels of maturity.[30] When the surveyed companies were asked: "Which explicit strategies do you follow in order to improve environmental performance?" interestingly, 38% of respondents deployed lean as a strategy for green. However, at the same time, corrective strategies such as waste reduction and waste recycling had a much higher adoption ratio, both at 83%. Overall, the findings revealed that, in a snapshot, companies were more likely to have end-of-pipe and corrective green initiatives, such as waste recycling and emissions offsetting, than proactive solutions, such as product design and supply chain optimization, confirming that most companies are still in less mature stages of integrating lean and green. It confirms that it is far more likely to find companies in a state of benign relationship between lean and green as opposed to synergistic or symbiotic relationships (Figure 3.9).

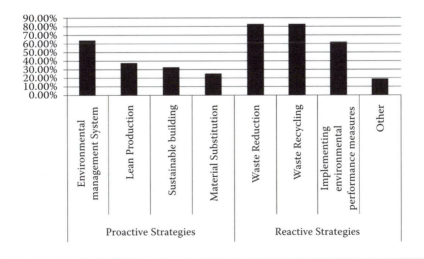

Figure 3.9   Carbon reduction strategies. (From K. Zokaei et al., *Best Practice Tools and Techniques for Carbon Reduction & Climate Change,* **Cardiff: Produced on behalf of CO$_2$ Sense Yorkshire, 2010.)**

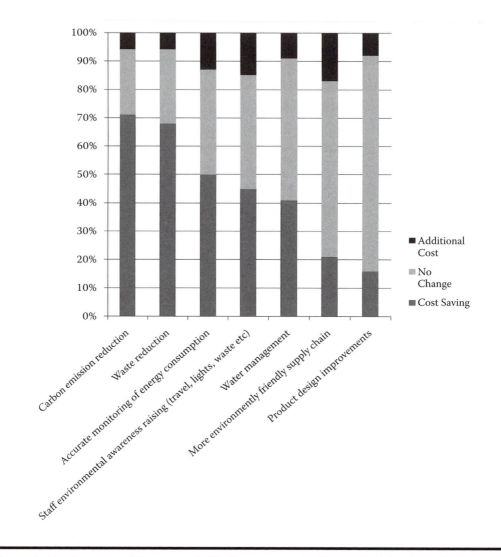

Figure 3.10 Impact of the green initiatives on costs. (From K. Zokaei et al., *Best Practice Tools and Techniques for Carbon Reduction & Climate Change*, Cardiff: Produced on behalf of CO$_2$ Sense Yorkshire, 2010.)

The surveyed companies were also asked to give feedback on the impacts of each one of their green initiatives in terms of economic cost reduction. As shown in Figure 3.10, carbon emissions reduction and waste reduction were highest ranked in terms of the respondents opinion about the associated cost savings while product design improvement and more environmentally friendly supply chain ranked low. Yet again, this is an indication of the *lack of awareness* about the benefits that more proactive initiatives can offer both environmentally and economically.

In terms of the users' perception of the effectiveness of various approaches in application, the results indicated that lean tools were ranked to have the highest level of effectiveness. But even more importantly, the large majority (i.e., 73%) of those who drew upon lean techniques for going green regarded cost as a strong driver behind implementing green initiatives. This demonstrates the importance of lean (both as a concept and as a toolbox) for simultaneous economic and environmental performance enhancement.

In summary, the survey showed that certainly lean is regarded by many managers as the means of green. However, while many firms are deploying lean tools and techniques for green, there is a lack of an integrative framework for creating a strategic symbiosis between lean and green. The lean community can play a key role in moving their companies beyond corrective approaches such as waste recycling and end-of-pipe solutions such as emissions capturing. In order to reap the real benefits, green initiatives should be systematically integrated into continuous improvement programs, and companies will need to adopt more proactive approaches such as whole life cycle thinking and lean and green product design.

Endnotes

1. J. Krafcik, "Comparative Analysis of Performance Indicators at World Auto Assembly Plants," Dissertation, MIT, 1988.
2. J. Womack, D. T. Jones, and D. Roos, *The Machine That Changed the World* (New York: Rawson Associates, 1990).
3. Ibid.
4. J. P. Womack, "The Power of Purpose, Process, and People," LEI Webinar, 1st May 1, 2008, http://www.Lean.org/Events/WebinarHome.cfm (accessed January 5, 2008).
5. W. E. Deming popularized the Plan-Do-Check-Act improvement cycle in post-World War II Japan. He also authored *Out of the Crisis* (Cambridge, MA: MIT Press, 1982).
6. J. Seddon, *Freedom from Command and Control* (Buckingham, UK: Vanguard Consulting Ltd., 2003).
7. G. Taguchi, *Introduction to Quality Engineering: Designing Quality into Products and Processes* (Tokyo: Asian Productivity Organization, 1986).
8. We are indebted to Bill Bellows, a quality guru at Pratt and Whitney Rocketdyne, who gives a full account of the story in the *Lean Management Journal* 1, no. 1 (2009): 7–11.
9. A term invented by John Seddon. See Seddon, *Freedom from Command and Control*.

10. "Relations with Employees," Toyota, accessed July 12, 2012, http://www.toyota-global.com/sustainability/stakeholders/employees/.

11. J. Womack, "Is Lean Green?" Lean Enterprise Institute, November 4, 2003, http://www.Lean.org/womack/ColumnArchive.cfm?y=2001&ey=2003#Col714.

12. Taiichi Ohno, *The Toyota Production System: Beyond Large-Scale Production.* (Cambridge, MA: Productivity Press, 1988, p. 75).

13. H. Daly, *Beyond Growth*, (Boston: Beacon Press Books, 1996).

14. D. Ehrenfeld, *The Arrogance of Humanism* (Oxford: Oxford University Press, 1981).

15. R. T. Watson and the Core Writing Team, eds., *Climate Change 2001: Synthesis Report. A Contribution of Working Groups I, II, and III to the Third Assessment Report of the Intergovernmental Panel on Climate Change* (Cambridge: Cambridge University Press, 2001).

16. R. K. Pachauri and A. Reisinger, eds., *Climate Change 2007: Synthesis Report, Contribution of Working Groups I, II and III to the Fourth Assessment Report of the Intergovernmental Panel on Climate Change* (Geneva, Switzerland: IPCC, 2007).

17. Ibid.

18. D. H. Meadows, D. L. Meadows, and J. Randers, J., *The Limits to Growth: A Report for the Club of Rome's Project on the Predicament of Mankind* (London: Pan, 1972).

19. *Our Common Future*, also known as the Brundtland report; World Commission on Environment and Development (WCED), Our Common Future (Oxford: Oxford University Press, 1987).

20. N. Stern, *The Economics of Climate Change: The Stern Review*, Cabinet Office (London: HM Treasury, 2006).

21. J. Elkington, *Cannibals with Forks: The Triple Bottom Line of 21st Century Business* (Stoney Creek, CT: New Society Publishers, 1998).

22. P. Hawken, A. B. Lovins, and L. H. Lovins, *Natural Capitalism* (London: Earthscan Publications Ltd., 1999).

23. E. U. von Weizsacker, *Factor Four: Doubling Wealth, Halving Resource Use, A Report to the Club of Rome* (London: Earthscan Ltd., 1998).

24. S. Schmidheiny, *Changing Course: Global Business Perspective on Development and the Environment* (Cambridge, MA: MIT Press, 1992).

25. W. McDonough and M. Braungart, *Cradle to Cradle: Remaking the Way We Make Things* (New York: North Point Press, 2002).

26. K.-H. Robert, *The Natural Step Story* (Gabriola Island, BC: New Society Publishers, 2002).

27. Kofi Annan, Speech before United Nations. Johannesburg, South Africa, September 1, 2002.

28. Sir Nicholas Stern, *The Economics of Climate Change.* Cambridge University Press, (Cambridge: England, 2007).

29. The Rio+20 United Nations Conference on Sustainable Development was held in Rio de Janeiro in June 2012. The outcome document was titled "The Future We Want" and it is available at: http://www.uncsd2012.org/rio20/thefuturewewant.html.

30. K. Zokaei et al., *Best Practice Tools and Techniques for Carbon Reduction & Climate Change* (Cardiff: Produced on behalf of CO_2 Sense Yorkshire, 2010).

CREATING A
LEAN AND GREEN
BUSINESS SYSTEM

Chapter 4

Lean and Green Business Process Management

In previous chapters we discussed about the shortage of systematic improvement approaches for integrating green and sustainability into business processes and offered the Lean & Green Business System model (Figure 4.1) as a structured and holistic means of greening businesses. As mentioned, the four sections of the model are the pieces of a jigsaw puzzle and you only see the whole picture once you get all of the pieces in the right place at the right time.

In this chapter we focus on the *process management* section of the model. In fact, historically, practitioners associate *lean* and *green* with operational improvements or with the tools and techniques used to deliver operational improvements. Although process management is probably the most critical element within the model, we know that lean and green process management per se is not going to deliver the full outcomes desired for a lean and green enterprise and that all areas of the model must be pursued simultaneously.

In the following, we will first illustrate a suite of tools and techniques developed for lean and green process management. Our lean and green toolkit is, by and large, an adaptation of lean techniques such as *value stream management* and *A3 management*. This is the most basic level of lean and green where companies essentially apply lean techniques for elimination of green wastes. The following illustrates that even this basic application of lean for green delivers significant improvements. One of us has been developing and applying the lean and green toolkit in different companies observing between 10% to 30% reduction in utilities as a result of rapid

Figure 4.1 Lean and Green Business System model.

kaizen blitz–type applications. Opportunities are enormous. This shows the value of having systematic approaches for greening operations and how fairly simple, yet proactive, methods can make a vast difference.

Subsequently, this chapter presents a case study of lean and green in the Toyota Motor Company. Toyota is, of course, the epitome of leanness; we investigate whether Toyota is green too. Our benchmarking illustrates Toyota's great lead over the rest of the industry and proves their immense commitment to social and environmental sustainability. Furthermore, we discuss a fresh term for lean and the Toyota Way, which so far remains largely unfamiliar in the Western literature on lean thinking and Toyota. This term is *Monozukuri*, which at the same time means sustainability. We discuss how the philosophy of Monozukuri—meaning manufacturing that is in harmony with nature and value adding for society—underlies the Toyota Way and the Toyota Production System.

Suite of Tools for Lean and Green Process Management

You will remember that we looked at the eight wastes of lean and put forward the eight wastes of green in Chapter 3. To remind ourselves, the eight wastes of green were excessive energy usage, physical waste (solid or liquid), excessive water usage, air emissions (most importantly the greenhouse gases), land contamination, discharges to water and effluents, noise and nuisance, and lost people potential. The aim of the lean and green toolkit ultimately is to eliminate these wastes from our processes.

Although the lean wastes and the green wastes have very different scope and nature, both sets are conceptualized for increasing the efficiency and effectiveness of the process or the *value stream* as we call it in lean terminology. In order to carry out improvements or waste elimination, a traditional lean thinker would first seek to map out the process and to identify where we should focus our efforts. After many years of working with lean value stream mapping techniques, we have found that there is no such thing as the "right" tool for mapping. It takes a combination of diagnostic approaches, all with their strengths and weaknesses. That is exactly why lean thinkers often use a combination of mapping techniques.

Figure 4.2 illustrates a typical lean diagnostic toolkit offering a range of value stream mapping techniques. This illustration also demonstrates the potential useful scope of each tool. For example, a common technique, such as big picture mapping, is useful in gaining an overview of the process and in looking into the flow of physical goods as well as the information flows. Nonetheless, it's not a very helpful tool when it comes to getting into the fine detail.[1]

In a similar fashion, we have developed the lean and green diagnostic toolkit, offering a range of mapping techniques aimed at analyzing and eliminating the green wastes as illustrated in Figure 4.3. Again, the illustration highlights the scope and the usefulness of each technique. Clearly, it is not the techniques per se that deliver benefits, but crucially the systematic intervention during which they are deployed. We provide a few examples of how they can be applied.

We have applied this toolkit to several companies of various sizes from very small to very large. The intervention is akin to an action learning Kaizen blitz method where the lean and green facilitator assembles a group of experts from different functions and works with them for 6 to 10 days over a period of a few weeks (sometimes just one full week). The aim is to deliver blitz improvements through application of the toolkit. The intervention is very hands-on and most workshops are held on the shop floor within the safety boundaries. The key characteristic of any Kaizen blitz intervention is that the team follows the plan-do-check-act (PDCA) cycle and spends lots of time on planning before doing.

Need	Lean Diagnostic Toolkit								
	Product Family Analysis	Process Decomposition	Big Picture Mapping	Process Activity Map	Four Fields Map	Supply Chain Response Matrix	Product Variety Funnel	Quality Filter Map	Demand Amplification Map
Gaining an overview	●	●	●			●			
Deciding where to start	●	●							
Scoping the process		●	●		●				
Working with physical products	●		●	●		●	●	●	●
Working with information	●		●		●				
Getting into fine detail				●	●		●	●	●
Working across organizations			●			●			

Figure 4.2 Lean diagnostic toolkit and scope of usefulness of value stream mapping techniques.

Business Need	System Boundary Mapping	Green Impact Matrix	Green Big Picture Map	Eco-Maps			Value Impact Ratio	Life Cycle Assessment	Footprint Analysis
				Water	Energy	Waste			
Gain overview	X	X	X				X	X	
Decide where to start	X	X	X				X	X	
Scope the process	X	X	X						
Walking the flows				X	X	X			X
Getting into fine detail				X	X	X		X	X
Working across organizations							X	X	X
Significant opportunity for innovation		X		X	X	X			

Figure 4.3 Lean and green diagnostic toolkit and scope of usefulness of mapping techniques.

In the first technique, the action learning team creates a Green System Boundary map, which is one of the simpler tools in our toolkit. This tool has been used to great effect by several companies that we have worked with, as illustrated in Figure 4.4. Simply put, the Green System Boundary map is a high-level mass and energy balance for the whole plant, including all energy and material going in and all energy and material coming out, even emissions and effluents. It would be even more powerful to include the cost of the inputs and outputs. Some more advanced companies are capable of providing a fairly complete picture in a snapshot, whereas others may require some time for data collection. Therefore, having templates or using an example such as the one in Figure 4.4 is useful. We have observed that, in spite of its deceptive simplicity, this technique can be very powerful in bringing about deep "realization moments."

What we are doing with this tool is looking at the balance of how much material is input into the firm (in terms of raw material, energy, and water), how much waste (gaseous, liquid, and solid) is produced, and consequently how much good product is made. A simple calculation can then show what percentage of good product is made by weight. In the example in Figure 4.5, 219 kg of input produces 143 kg of waste and 76 kg of good product, a system efficiency of 34.7%. This Current State map may be used as a basis of a Future State map, where this can be increased to maybe 50% or more.

Emissions 22 T CO2-e	

| Gas | 400,000 m3 |
| Cost | $250,000 |

| Electricity | 3,000,000 KWh |
| Cost | $350,000 |

| Water | 24,000 m3 |
| Cost | $57,000 |

Materials	
Packaging etc	1500T
Raw Product	700T
Cooked Product	1800T
Other Product	400T
TOTAL	4,400 Tonnes

Finished Product	
Total	3,800 Tonnes

Waste	
Raw Waste	40T
Rubbish	270T
Total waste	310T
Cost	$55,000

Recyclables	
Cardboard	350T
Plastic	12T
Oil drums	1T
Total	363 T
Income	$15,000
Equip. Rental	$10,000
Surplus	$5,000

| Effluent | not measured |
| Value | $120,000 |

Figure 4.4   Green System Boundary map.

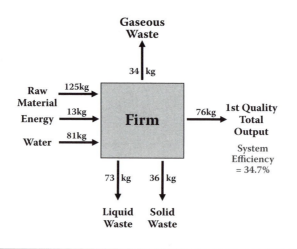

Figure 4.5   Green System Boundary map and system efficiency.

Clearly, increasing the efficiency will deliver the dual goals of resource efficiency (green) and economy (lean) at the same time.

The second technique that the action learning teams apply is the Green Impact Matrix, which is illustrated in Figure 4.6. The Green Impact Matrix is a rapid solution for identifying the key hotspots of the green waste along different stages of the process. In this matrix, the eight wastes of green are along the vertical axis and different stages of the process are along the top. The action learning team populates the matrix with color-coded boxes based on consensus. For example, for energy usage, the team will look at the level of energy intensity at each stage of the process where green is low intensity (for example, less than 3% of plant's total energy), amber is medium usage (say, between 3% and 15%), and red is high intensity (more than 15%). The

Green Impact Matrix								
Green Wastes	Process 1	Process 2	Process 3	Process 4	Process 5	Process 6	Process 7	Process 8
Energy Consumption								
Air Emissions								
Water Usage								
Physical Waste								
Land Contamination								
Discharge to Water								
Noise and Nuisance								
Lost People Potential								
MAP KEY	Impact elsewhere	**Blue**	Hi Impact	**Red**	Med Impact	**Amber**	Lo Impact	**Green**

Figure 4.6   The Green Impact Matrix.

point is applying the technique as a quick scan in an action learning workshop rather than requiring the team to glean detailed evidence for color coding the process. More detailed analysis can follow later.

Having completed the Green Impact Matrix, the team will be able to tell which areas are suitable for further investigation and action. For example, in Figure 4.6, process 1 has many red codes while process 8 has none. This mapping technique often sparks healthy discussions among team members leading to the need for further investigation, which can be followed up during the gaps the facilitators design between the blitz workshops for the purpose of data gathering. Again, while this technique is very simple, its application in an action learning environment can be very rewarding.

In the rest of the Kaizen workshops, the project team creates a Green Big Picture map as illustrated in Figure 4.7. In this example the team selected five hotspots from the Green Impact Matrix for improvement. Each hotspot was assigned an "Eco-map," which is the next tool in the toolbox. Different team members took the lead in completing the Eco-maps for each of the five selected hotspots.

The Eco-maps were essentially the same as *A3 problem solving*, which in this case was deployed for green improvements. A3 problem solving is a lean technique and a management concept used by Toyota and by many lean practitioners. It provides a common language for problem solving across the organization. It is simple and powerful. It is just an A3-size piece of paper where the left half is dedicated to problem investigation (plan) and the right half is dedicated to solution and sustainment of the solution (do, check, and act). There is a project owner, a senior sponsor, and various stakeholders clearly named on the

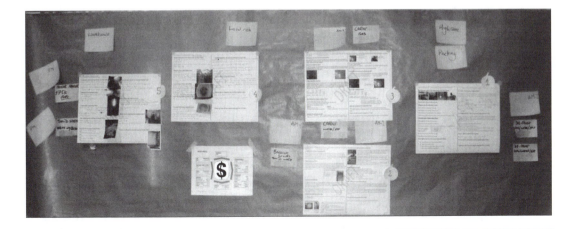

Figure 4.7 Green Big Picture map.

A3. We suggest that you refer to various sources on A3 management if you are not sure about the tool or if you have not seen one in action.

During the final stage of the action learning lean and green intervention, the team members create a detailed action plan for each of the selected hotspots and propose how to eliminate the green waste. When applying the toolkit, you should aim to generate an engaging environment to capture and develop innovative ideas and to take more ownership which lead to incredible results. Each one of the A3s has its own dedicated local team that is supported by the project facilitator and senior executives.

The A3 technique is highly visual and begins by going to the workplace and observing the waste issue. The A3 owner needs to work closely with the respective operator, or even better, be the process owner. Once the A3 counter measures are generated (typically initial solutions take one to two weeks), the facilitators collect all suggested opportunities for a final presentation and additional approval from the senior sponsors. This is an opportunity to quantify the total size of the prize. In our experience, between 10% and 30% of the utility footprint and cost can be reduced by applying this lean and green Kaizen blitz approach.

In one example from a multinational food manufacturer, various A3 projects led to identification of solutions to reduce the energy usage by 16% and the water footprint by 15%. This was estimated to be equivalent to around $1.2 million utility cost savings (based on the UK energy and water prices at the time). The identified solutions were based on very low-investment approaches, where the team stayed within their own budgetary authority. This means that we did not require additional sign-off approval. This shows the value of having a systematic approach for greening operations and demonstrates how relatively simple, yet proactive, methods can make a vast difference if applied in a methodical way.

We are not going into further details about the application of the rest of the lean and green toolkit. It is enough to say that the power of the lean and green toolkit is in the action learning approach, which can harness the abilities of a wide range of stakeholders such as utility engineers, process engineers, operations managers, environmental managers, quality managers, and line operators. But more crucially, the results discussed in the previous paragraphs once again show the potential size of the prize in putting lean and green together. We have repeatedly worked with companies who have delivered considerable savings beyond their initial expectations, by applying the lean and green toolkit described here. Our suggestion is to retain some of the savings from the first lean and green kaizen for future green investments.

In the following we focus on Toyota as a prime example of "lean and green process management." Firstly, we begin with an interview with an executive from Toyota Motor Company. We then continue to provide a fuller account of Toyota's green initiatives.

* * *

Interview with Steve Hope

General Manager, Environmental Affairs and Corporate Citizenship, Toyota Motor Europe

Interviewer: *Can you tell us about yourself, your career experience, and how you came to be in your current role at Toyota?*

Steve Hope: *I graduated as an electrical engineer from Bradford University and started work with the Boots Company, working finally as the power and services manager working in the CHP power station. Subsequently I joined Toyota in 1991 as a facilities and environment manager. I was interested in this role because of my background as an electrical engineer in the facility role. At that time it was quite unusual, if not unique, for any engineer to have the term environment in their job title. This is how my story in environmental management began in Toyota where I spent the first 15 years at Toyota UK, in a variety of roles such as facilities and environment and engineering project management for buildings and new facilities. I was involved in construction projects for expansions in the UK and in France. I also had new experiences in areas such as assembly engineering and health and safety, which offered fresh learning. However, my focus included the environment, all the way through that time. In 2006, I was invited to join Toyota Motor Europe as Senior Manager, Facilities and Environment, working with all nine manufacturing sites across Europe. Subsequently I was promoted to the role of General Manager, which meant that my remit expanded to include Construction, and later Health and Safety as well as Facilities and Environment. As of January 2012, I was invited to change position again to become General Manager for Environmental Affairs and Corporate Citizenship. This change means that my scope is not just limited to the manufacturing portion of the business but also R&D, Logistics, After Sales Services (product use), and End of Life. At Toyota we adopt a full 360-degree or whole life-cycle perspective on our environmental impacts, so we consider the full scope of sustainability. When I first started at Toyota, the term* sustainability *was not even in use yet; however, the environmental impact assessment for the Burnaston plant in the UK was one of the first environmental assessments in the UK.*

Interviewer: *Why does Toyota care about environmental affairs and corporate citizenship?*

Steve Hope: *It's part of our ethos and philosophy. It goes right back to the guiding principles of Toyota. In 1992 the Toyota Guiding Principles were established, which provide the cornerstone of Toyota's corporate management philosophy and provide Toyota with a clear path for achieving sustainability.*

* * *

We have a role to exercise in relation to the society. We don't think we can manufacture what we want to manufacture independently of the impacts on the wider society. Our philosophy is that we are an integral part of the

GUIDING PRINCIPLES AT TOYOTA

1. Honor the language and spirit of the law of every nation and undertake open and fair business activities to be a good corporate citizen of the world.
2. Respect the culture and customs of every nation and contribute to economic and social development through corporate activities in the communities.
3. Dedicate business to providing clean and safe products and to enhancing the quality of life everywhere through all our activities.
4. Create and develop advanced technologies and provide outstanding products and services that fulfil the needs of customers worldwide.
5. Foster a corporate culture that enhances both individual creativity and the value of teamwork, while honoring mutual trust and respect between labor and management.
6. Pursue growth through harmony with the global community via innovative management.
7. Work with business partners in research and manufacture to achieve stable, long-term growth and mutual benefits, while keeping ourselves open to new partnerships.

environment and we are fulfilling a need of society, which is the desire for personal transportation.

In 1992 Toyota also published its Earth Charter. *This was established in direct response to the international initiatives agreed upon at the Rio Earth Summit, and through their application and revisions, continue to guide our reductions in our environmental impact.*

Interviewer: *What is Toyota's vision in terms of Corporate Citizenship of Tomorrow?*

TOYOTA EARTH CHARTER

IMPLEMENTING CONSOLIDATED ENVIRONMENTAL MANAGEMENT

Basic Policy

1. Contribution toward a prosperous 21st century society: Contribute toward a prosperous 21st century society. Aim for growth that is in harmony with the environment and set as a challenge the achievement of zero emissions throughout all areas of business activities.

2. Pursuit of environmental technologies: Pursue all possible environmental technologies, developing and establishing new technologies to enable the environment and economy to coexist harmoniously.

3. Voluntary actions: Develop a voluntary improvement plan, based on thorough preventive measures and compliance with laws, which addresses environmental issues on the global, national and regional scales and promotes continuous implementation.

4. Working in cooperation with society: Build close and cooperative relationships with a wide spectrum of individuals and organizations involved in environmental preservation including governments, local municipalities, with related companies and industries.

Action guidelines

1. Always be concerned about the environment: Take on the challenge of achieving zero emissions at all stages, i.e., production, utilization and disposal.

   (1) Develop and provide products with top-level environmental performance.

(2) Pursue production activities that do not generate waste.

(3) Implement thorough preventive measures.

(4) Promote businesses that contribute toward environmental improvement.

2. Business partners are partners in creating a better environment: Co-operate with associated companies.

3. As a member of society: Actively participate in social actions.

(1) Participate in creation of a recycling-based society.

(2) Support government environmental policies.

(3) Contribute also to non-profit activities.

4. Toward better understanding: Actively disclose information and promote environmental awareness.

Steve Hope: *As we see ourselves, as an integral part of the wider society, our role is not only to fulfill society's desire for transportation, but also beyond, to address the wider social concerns. We see it as our responsibility to work not just inside our own operations, but also with other partners, including our suppliers, fuel producers, governments, nongovernmental organizations, and customers to address a sustainable future for the society. These issues include traffic safety and congestion, education, and importance of culture in society in addition to the environment.*

Interviewer: *What are the main KPIs [key performance indicators] that you use to measure achieving your vision?*

Steve Hope: *We deploy clear measurement methods to understand environmental performance at all levels and to link performance from shop floor KPIs all the way to the board level. These measures are aligned and cascaded, and for each individual their scope depends on the level of that person's responsibility. We use measures such as CO_2 (linked to energy use), water usage, and waste to demonstrate to our members their respective progress against (often challenging) targets that we set. For example, at the moment, out of the 9 manufacturing plants across Europe, 8 are zero to landfill on the waste measure. The only plant that is not zero to landfill is in Russia, where the required recycling and waste management infrastructure is partly not available. Moreover, several plants are achieving zero waste to incineration (unless used for energy recovery). Under* *the EU [European Union] definitions, these plants are considered as operating at "zero disposal." It's key to have cascaded measurement systems, which provide effective KPIs at the right level to enable our people to improve continuously. Across all Toyota plants, there are basic measures in place for everybody, all the way from top board to the local level. These are, in the order of our priority: safety, environment, quality, production, and cost. We regard safety of our members and our customers together with our environmental commitment to society as our permission to operate. Improvement is continuously measured against these five criteria.*

Interviewer: *Is your corporate citizenship commitment also profitable?*

Steve Hope: *Yes. Often when we identify opportunities for environmental improvement, they also contribute to achieving the cost KPI. Sometimes there are also improvements in terms of safety, quality, and production. This is a good way to know a project's priority. Pursuit of environmental perfection also makes us become more innovative, to challenge the unknown. For example, when we were first faced with the challenge to achieve zero to landfill in our production plants, we did not know how we could achieve it, but we had given ourselves a 5-year target. This meant that we had to experiment with various solutions and engage our members, capturing their innovative ideas and rolling out best practices continuously. In the end we achieved this target 3 years ahead of schedule in the UK. In general, innovative organizations are more likely to be profitable. Similarly, making products more suitable for end-of-life recycling means that often we can also reduce cost or create additional revenues from recycling.*

Interviewer: *Can you explain the relationship between Toyota Production System [TPS] and your Corporate Citizenship commitments?*

Steve Hope: *TPS was born out of measures for relatively austere times. The history of the development of TPS is that we had limited resources where, for example, Toyota could not afford to commit large amounts of capital for machines or for work in progress. In the press shop, for example, Taiichi Ohno, the father of Toyota Production System, began manufacturing with few presses and smaller batches, which has today evolved into the ability for most press shops to achieve exchange of dies in under 10 minutes [single minutes exchange of dies]. From the beginning, Toyota has been pursuing the elimination of Muda [waste], Mura [unevenness], and Muri [overburden]. There are seven types of Muda, elimination of which has a direct impact on our environmental improvement. So there is a direct link between relentless application of the Toyota Production System and achieving greater social and environment benefits. There is a culture of continuous improvement within Toyota. We standardize and then apply Kaizen and then we standardize the improvements again. The same culture of continuous improvement is applied to achieve our environmental targets. When an environmental initiative is implemented, it is sometimes the case that costs can go up temporarily. However, we then apply Kaizen to bring cost down to levels even lower than before the initiative was implemented. For example, when the electric engine was fitted into the hybrid vehicle, Toyota—initially—had to increase the rare metals content in order to miniaturize the electric motor with the same efficiency levels. However, the next step was to dematerialize and to reduce the rare metals content in the magnet. This has been ongoing through the application of the Kaizen and continuous improvement, followed by rigorous standardization, which not only reduce cost but also lower the material impact of the vehicle.*

Adaptability or continuous learning is another key feature of TPS. We are flexible and aim to capture the latest environmental trends and information. For example, currently the focus of environmental measurements [is] spreading from energy use and greenhouse gas emissions to the use of materials, especially precious materials. While the external forces are shifting to capture the impact and the use of materials, we are actively engaging with the environmental community to address the issue. We are measuring and reducing our own precious material content as well as expanding our recycling activities. At the same time, we try to share some of our previous experiences with other industries.

Furthermore, while Kaizen is a great enabler for Toyota environmental targets, we know that there is the law of diminishing returns in applying continuous improvement, and therefore you also need to have a step change. We continuously apply Kaizen and occasionally we are able to apply a step change during plant refurbishment or a model change to achieve environmental improvements. When opportunities are identified, we apply our five KPI criteria in order to select the correct direction. As previously mentioned, these are, in the order of importance, safety, environment, quality, productivity, and cost.

All members within Toyota family have measures linked into these five criteria with clear targets to achieve improvements.

* * *

Lean and Green at Toyota Motor Company[2]

Toyota can easily be the most storied company in the world. For decades it has caught the imagination of managers around the world, becoming an emblematic case for quality, reliability, productivity, employee engagement, continuous improvement, cost reduction, expansion, and market capitalization growth. The Toyota Production System (TPS) is synonymous with efficiency, productivity, and customer satisfaction, and in spite of a few hiccups in recent years, it remains a source of inspiration for continuous improvement in every aspect of an enterprise.

Here we trace not only the high points of their legendary production system, but also the less documented and more recent developments around the convergence of lean and green thinking in Toyota. In the face of the pressing environmental and energy challenges, Toyota continuously refines their vision and practices in the quest to stay true to their core values and to remain as one of the most consistently successful companies on the planet. In the following, we benchmark various aspects of Toyota's environmental performance and explain how they outperform the competition on many levels.

Toyota's culture of continuous innovation, combined with their corporate values, is manifested in an everyday commitment to environmental excellence. There are numerous accounts of Toyota's remarkable achievements, but there has been a lack of attention to the fundamental reason for their success, which is in Toyota's spirit of seeing themselves as an integral part of the society serving the community's desire for transportation rather than simply being a money-making machine. Toyota's corporate spirit is passed on from the time of the founder to the present day, in the form of the *Toyoda Precepts*. The precepts capture the intentions and the thinking of the founder of the Toyota Group, Mr. Sakichi Toyoda, and form the values that the company lives to date.

> Initially this did not have a definite shape. However, with the growth of the scope of the company, the need arose for the principles to be codified so that they could be propagated among employees. Risaburo Toyoda and Kiichiro Toyoda, in the pioneer

days of the company, gathered together the teachings of Sakichi Toyoda and published them in the form of the Toyoda Precepts on October 30, 1935, the fifth anniversary of his death. From that time, the precepts have played the role of a spiritual support for employees as the principles of the company. And this spirit of the Toyoda Precepts can still be felt today.[3]

The precepts in turn shape the *Guiding Principles* of the Toyota Motor Corporation.

TOYODA PRECEPTS

1. Be contributive to the development and welfare of the country [later was expanded to give it a global focus] by working together, regardless of position, in faithfully fulfilling your duties.
2. Be ahead of the times through endless creativity, inquisitiveness and pursuit of improvement.
3. Be practical and avoid frivolity.
4. Be kind and generous; strive to create a warm, homelike atmosphere.
5. Be reverent, and show gratitude for things great and small in thought and deed ("Toyota," 2003).

A Brief History

When Sakichi Toyoda invented the wooden Toyota handloom in 1890 and later sold the patent rights to the British,[4] he did not imagine that his continuous curiosity in powered engines would inspire his son, Kiichiro, and his nephew, Eiji Toyoda, to establish the most influential production systems in the contemporary world.[5] Inspired by the work of Eli Whitney, one of the originators of interchangeable parts, and by Frederick Taylor's *Scientific Management*, Kiichiro traveled to the United States in the 1930s[6] to observe the mass production system at Ford. This led to the launch of the company's first production line—the AA passenger prototype—in 1936, using a synchronized system to produce 20,000 vehicles, essentially setting the foundation of the *just-in-time* system at the present Toyota Motor Company.[7]

However, it was Taiichi Ohno who really instituted the Toyota Production System as we know it today. Indeed, Ohno had spotted two fundamental flaws in the Western production systems: first, large inventories that tie up capital and hide quality defects, and second, the inability to accommodate consumer preferences for product diversity.[8] From 1948 onwards, Ohno gradually extended his concept of small-lot production throughout Toyota, which started in the engine machining shop to create the TPS. Ohno was inspired by Kiichiro's reasoning that the best way to work would be to have all the parts for assembly at the side of the line *just in time* for the user.[9] In order to facilitate the just-in-time flow of parts in production, Ohno defined the seven wastes to be eliminated: *overproduction, defects, unnecessary inventory, unnecessary transport, waiting (idle people and machines), inappropriate processing,* and *unnecessary motion.* Later on, other lean thinkers added an eighth waste, which is *wasted human potential.* We already discussed the eight wastes of lean in Chapter 3. The aim of eliminating the lean wastes was to create a better flow rather than just reducing *activity cost.* This is a point that many Western companies missed. Many Western managers perceive TPS as a continuous method for waste elimination where waste is related to activities on the job. However, Ohno knew very well that the *real cost is not in activities but rather in interrupting the process flow.* Ohno modified the machine changeover procedures to produce a growing variety in smaller lot sizes in considerably reduced lead times. The result was an ability to produce a considerable variety of automobiles in comparatively low volumes at a competitive cost, challenging the conventional logic of mass production.

Toyota defines TPS as "a production system which is steeped in the philosophy of "the complete elimination of all waste" imbuing all aspects of production in pursuit of the most efficient methods."[10] According to Ohno (1988), TPS has two pillars where *just-in-time flow* is only one. The other pillar is *Jidoka.*[11] Jidoka can be translated as *automation with a human touch, intelligent automation,* or *autonomation,* which means when a problem occurs, production stops immediately, preventing production of defective products, until a lasting fix is put into place and root causes are addressed. The point of Jidoka is to stop the line to fix the problem and not to allow the quality defect to pass on to the next station.

On the surface, this principle even appears to be disruptive to just-in-time flow. However, Toyota insists on stopping the line for quality when needed, in order to instill a quality culture of *right first time every time* and to implement the *100-year fix*, avoiding recurring defects even if it means stopping the line, interrupting the flow, and costing money. With Jidoka, the worker stops the process to ensure that quality is perfect at any cost. It means that frontline employees are empowered and enabled to deliver perfect quality. The frontline worker will pull a cord or otherwise send a signal for the line to stop when he or she spots a quality defect. Ohno developed the notion of Jidoka, observing Sakichi Toyoda's automatic loom, which stopped once the thread broke, disallowing defective fabric to be woven and avoiding wasted material and machine time.[12]

Far from turning the worker into the adjunct of the machine, TPS insists on frontline employees' ability to make the key decisions, to absorb variation, and to react to quality issues as much as possible. Whereas simple automation is concerned with efficiency and labor reduction, *autonomation* focuses on effectiveness, on quality improvement, and even more importantly on the priority of the worker over the machine or the system. In Toyota's language, automation equally applies to any sort of systemization—any situation that involves replacing human activities or decision making with a machine. As such, systems and machines are always subordinated to humans.

Simply put, Jidoka is about building quality into the process by putting thinking and authority back into frontline working. That's why in the TPS, standardization is a way of workers helping themselves to improve the process, rather than a method of control. Here is a fundamental difference with the Western method of management where systemization is a method for controlling people.

Compare this to any average manufacturing process or even call center where frontline staff have very little authority to address variation in work outside the imposed controls, let alone to stop the line and ask for help. Today, dumbed-down systemization is ubiquitous in both manufacturing and service in Western companies. You may experience such dumbed-down standardization when calling a help desk or a financial services call center. You get call handlers who are unable to deal with your issue effectively, that is if you actually talk to a human after futilely dealing with an automated voice recognition machine for several minutes. In manufacturing, where frontline staff are hardly ever encouraged to think, the situation is not much better; a manufacturing manager once told one of us "I only employ their limbs, not their brains," referring to the company's frontline staff.

Ohno and his disciples built the TPS as it stands today based on a solid understanding of systems principles and a deep appreciation of the teachings of Dr. Deming and other early quality gurus discussed in Chapter 3. To them, most important were the core ideas, the precepts of the founding

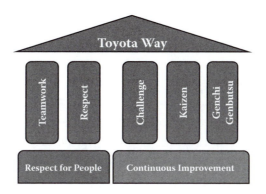

Figure 4.8   The Principles of the Toyota Way.

fathers, which live through the company to date—core ideas that revolve around having a sense of duty to contribute to the development and the welfare of society at large, rather than seeing the company just as a money-making machine. Compare this with the present day banking culture!

In 2001 Toyota articulated their corporate philosophy and their approach in a document called the *Toyota Way*.[13] The Toyota Way (Figure 4.8) is different from, and should not be confused with, the Toyota Production System. It encompasses more than just production, covering all aspects of the enterprise from leadership and human relations to strategy and production. It is an expression of the company values as well as a set of guidelines that all employees should embrace in all aspects of their jobs in any function. Formed on the twin pillars of *continuous improvement* and *respect for people*, the Toyota Way consists of the following five principles:

■ **Challenge:** Form a long-term vision, meet challenges with courage and creativity to realize the vision.
■ **Kaizen:** This is Japanese for improvement or change for better. Improve the business continuously, always driving for innovation.
■ **Genchi Genbutsu:** This is Japanese and it means go and see. It is a key principle of the Toyota Way. It means going to the source or to the workplace to find facts and make the right decisions. Take the decision making as close to the workplace as possible.
■ **Respect:** Respect others by understanding them and try best to build trust. Understanding only comes through a deep appreciation of how work works and it is possible by getting closer and closer to the workplace. Command-and-control managers are unable to truly respect others because they are far removed from the frontline.

■ **Teamwork:** Drive personal and team growth. Share opportunities and be collaborative to maximize individual and team performance.

Learning from the Toyota Way

As Toyota began to expand beyond its home base during the 1980s, the Toyota Way and its transplantation to other Western companies became the focus of investigation of many researchers intending to unravel the secrets of Toyota's remarkable performance. One of the first research consortia looking at the competitive performance of the automobile manufacturers was the International Motor Vehicle Programme (IMVP), which started in 1980. A key output of IMVP was the publication of *The Machine That Changed the World*[14] in 1990, which drew attention to Toyota's immense efficiency and quality lead over Western carmakers such as GM and Ford. Table 4.1, from *The Machine That Changed the World*, shows that Toyota took only half as long as GM did to manufacture a similar size vehicle, with a fraction of the number of defects.

The authors of *The Machine that Changed the World* used the phrase *lean production* to conceptualize Toyota's ability to do more with less; lean implied more productivity, better quality, and less inventory[15] than conventional mass production in the West. Ever since, lean has become synonymous with the Toyota Production System and the Toyota Way. The lean movement has helped Western manufacturers to bridge the gap with Toyota to a large extent, at least in terms of the more obvious measures

Table 4.1 General Motors versus Toyota versus NUMMI in 1987

	GM Framingham, MA, USA	NUMMI (Joint Venture between Toyota and GM in Fremont, CA, USA)	Toyota Takaoka, Japan
Assembly Productivity Hours/car	31	19	16
Assembly Quality Defects/car	135	45	45
Average Inventory of Parts (measure of delivery)	2 weeks	2 days	2 hours

Source: J. P. Womack, D. T. Jones, and D. Roos, *The Machine That Changed the World* (New York: Simon and Schuster, 1990), 83.

of productivity and quality. In the United States, Detroit's Big Three have improved their productivity according to an industry benchmark. In 1995, a GM car took on average 46 hours to make, Chrysler 43, and Toyota only 29.4. By 2007, GM had improved productivity to 32.3 hours per vehicle and Chrysler 30.4, while Toyota remained almost static at 29.9 hours per vehicle.[16] Today, you scarcely find a large corporation that doesn't have a vibrant lean, continuous improvement, or a similar initiative. Even small and medium-sized enterprises are learning from Toyota to optimize their processes. One may argue, though, that the lean movement, at least so far, has not really managed to transfer the fundamental precepts underlying Toyota's remarkable performance.

Toyota's Environmental Performance

Responsible for 14% of all greenhouse gas emissions worldwide,[17] the transport industry has been pressed more than any other sector of the economy to shift toward Kyoto Protocol target levels. It is estimated that automobiles, worldwide, emit around 10% of global emissions from fossil fuels and 6% of all global warming potential.[18] Just to make the scale of the problem clear, in the United States, the yearly carbon emissions from only GM-manufactured vehicles on the road are more than double those of America's largest electric generating company, that is, American Electric Power. In fact, the CO_2 emissions of each of Detroit's Big Three automakers exceed those of American Electric Power.[19]

Thus, it is not surprising that there are legislative, social, and political pressures on vehicle manufacturers to reduce their global warming contribution both during production and in use. In recent years, most volume car manufacturers have set fairly ambitious targets to reduce their emissions. Early integration of environmental concerns into the day-to-day operations of the company has helped Toyota establish itself as an environmental leader in the field. Toyota has been consistently reducing its environmental impacts, which in turn not only has facilitated compliance with emerging environmental regulations in its main markets of the United States, Japan, and Europe, but also has helped Toyota gain considerable competitive advantage in many respects.

Here we benchmark Toyota's environmental performance against the industry from different perspectives. It is not easy to measure any given company's environmental performance, especially in an industry as complex as the automotive industry. One can look at environmental performance from many different aspects, from global warming potential to biodiversity and from waste management to public health. Surely, all aspects are equally important and no single one can be discounted. Moreover, unlike the economic currencies, "environmental currencies" for measuring performance are varied and uncertain. Nevertheless, since the automotive sector is a key contributor to global warming, it is generally accepted that reduction in greenhouse gas (GHG) emissions is a key issue for car makers and CO_2-equivalent is a meaningful enough "currency."

However, even benchmarking against GHG emissions alone has its own complexities. First, it is essential to understand what part of the product life cycle is being analyzed. For example, as we show in Figure 4.9, there are various sources of GHG emissions during the end-to-end life cycle of an automobile.

A study at the University of Michigan showed that nearly 85% of the emissions from the end-to-end life cycle of an average passenger car can be attributed to emissions during use.[20] Of course, this emphasizes the criticality of manufacturing vehicles that have better fuel economy and emit fewer grams of GHGs per mile. Toyota is already a leader in launching cleaner cars through their hybrid and plug-in hybrid technologies, most notably the Prius.

However, comparing automakers by taking into account the entire life cycle of all their vehicles ever produced is extremely complex. Here we provide an important benchmark for measuring the performance of Toyota against the rest of the industry in terms of its *total emissions* (direct and indirect) highlighted in Figure 4.9.[21] Having benchmarked Toyota against other automakers in terms of the total emissions, we will now look at Toyota's strategy for reducing their vehicle emissions during use.

Figure 4.10 shows *total emissions* from a number of major volume car manufacturers between 2008 and 2011. It is evident that Toyota has been consistently reducing their total emissions. This trend is consistent with various pledges made by Toyota to reduce their volume of GHG emissions in all production and related areas. For example, Toyota set for itself a worldwide target to reduce emissions per unit of sales by 20% by 2010 against 2001 levels. They actually delivered a 23% reduction from 2001 levels and a 51% reduction from 1990.[22]

These improvements are rooted in Toyota's Kaizen (continuous improvement) mentality. The day-to-day Kaizen in Toyota is guided and driven by five-year action plans, which set tangible goals for all areas including production, facilities, transport, and offices as well as specific regional targets for various plants. These targets cover areas such as energy, emissions,

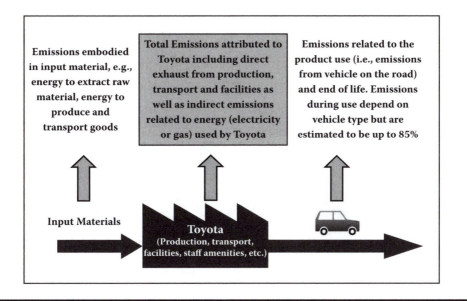

Figure 4.9   There are various sources of GHG emissions during the end-to-end life cycle of an automobile.

water, waste management, hazardous substances, and atmospheric quality. The targets are not separate from employees' day-to-day operation, but rather are integrated into the company's *strategy deployment* mechanism or *Hoshin* planning. Through Hoshin planning, Toyota aligns the top and the bottom of the organization. Toyota believes that managers must delegate as much authority as possible. "That is the way to establish respect for humanity as your management philosophy. It is a management system in which all employees participate, from the top down and from the bottom up, and humanity is fully respected."[23] Through rigorous Hoshin planning, Toyota ensures that environmental goals are integrated into everyone's working life.

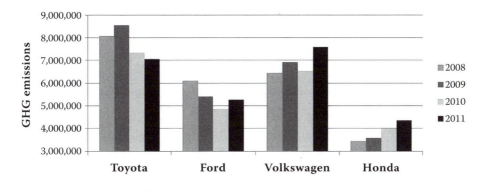

Figure 4.10   Total GHG emissions from selected volume car makers.

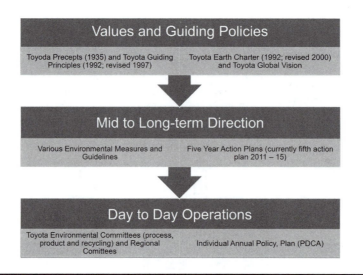

Figure 4.11   Toyota Environmental Action Plan System. Figure adapted from Toyota Motor Corporation (2011) Toyota Environmental Report, [online] www.toyota.no/ Images/environmental_report11_fe_tcm308-1100813.pdf

Moreover, because Hoshin planning is so deeply rooted in the notion of *respect*, Toyota members often exceed their targets and deliver results that are much more sustainable than any organization where targets are enacted by means of control or even bonuses.

As illustrated in Figure 4.11, the annual plans and goals at Toyota are linked to the five-year action plan, which is in turn rooted in the Toyota Earth Charter and Toyota Global Vision. The Toyota Earth Charter itself is based on the Toyota Guiding principles, first codified in 1992. There are a number of Toyota Environment Committees covering product design, production, and resource utilization, which oversee the implementation and adherence to the environmental five-year action plans.[24] Currently, Toyota is in its fifth global Environmental Five-Year Action Plan (2011–2015). There is a strong and successful history of implementing environmental initiatives within Toyota, with the establishment of the first Toyota Environment Committee dating as far back as 1963. Moreover, there is the main Environment Committee chaired by company president, Mr. Akio Toyoda, which directs Toyota's priority implementation issues. This is to ensure that ongoing commitment to environment is in place at the highest possible level.

Therefore, environmental planning and execution is not just culture that Toyota inherits from its founding fathers. It is rather a rigorous process of planning, refining, and implementing, which is an integral part of the company's strategy deployment or Hoshin. Culture is something that Toyota creates with conscious devotion and nurtures through zealous care. With nearly 320,000 employees around the globe, and with hundreds of sites to manage, Toyota faces the same resistance and skepticisms as any other company, if not more. What sets Toyota apart is their unwavering commitment to a vision of being "kind to Earth" and having "respect for all people."[25]

So far we have looked at Toyota's total emissions in terms of the absolute amount of the annual GHGs. Although this provides a relatively useful measure for understating trends (increase or decrease in GHGs) for the same company, it is hardly useful for comparing among companies. For example, in Figure 4.10, Honda's emissions cannot be compared to Toyota's since the production volumes and turnovers are quite different. Therefore, in Table 4.2 we attempt to provide a more meaningful way of benchmarking. We use three different measures to explain total GHG emissions in terms of the number of vehicles manufactured, turnover, and number of staff. As far as we know, this is the first time such a benchmark is being made available. We use the latest available data from 2010 triangulating information from a number of different sources to arrive at the best possible benchmark.

Toyota tops the list in terms of the lowest *tons of CO_2 emitted per vehicle manufactured* which is arguably the most reliable measure for benchmarking environmental performance across the sector. Toyota also tops the ranking in terms of *tons of CO_2 emitted per US\$1 million in revenue*, if we discount Daimler (Mercedes), which predominantly assembles luxury vehicles sold at premium prices. In this category, Toyota is closely rivaled by Honda and Fiat (the latter of which also owns luxury brands such as Ferrari, Maserati, and Alfa Romeo). Last, but not least, we benchmarked Toyota in terms of *total emissions per employee*. Again, Toyota is performing significantly better than its close rivals, GM and Ford, but it is behind Volkswagen, which employs around 20% more staff. This can be explained partly due to differences in employment policies and partly due to Toyota's higher productivity performance per staff member, especially in Japan.

All in all, Toyota demonstrates excellent environmental performance compared against the rest of the industry. One could liken this to the original industry benchmarks by the IMVP researchers that first highlighted Toyota's superiority on key economic measures (Table 4.1). Over the last few years, companies such as General Motors have worked to bridge their environmental gap against Toyota and, at least to some extent, they have been successful.

The commitment to reduce costs while searching for sustainable profit margins over time has permeated Toyota. Since the 1930s, Toyota has

Table 4.2 Benchmarking Toyota Emissions versus Competitors

Year 2010	Number of Staff	Revenue in $M	Number of Vehicles Manufactured	Ton CO_2-e	T CO_2/Vehicle Manufactured	T CO_2/$M Sale	T CO_2/ Staff
Toyota	320,808	222,000	8,557,351	7,334,000	0.86	33.0	22.9
GM	209,000	135,592	8,476,192	7,863,406	0.93	58.0	37.6
Fiat S.P.A.	190,014	72,200	2,716,286	2,663,645	0.98	36.9	14.0
Volkswagen	399,381	162,851	7,341,065	7,700,000	1.05	47.3	19.3
Ford	164,000	128,954	4,988,031	5,300,000	1.06	41.1	32.3
Honda	181,876	120,270	3,643,057	4,000,000	1.10	33.3	22.0
Daimler (Mercedes)	260,100	130,900	2,410,021	3,699,102	1.53	28.3	14.1

Source: Company Annual Reports and Ecodesk[26] | Ecodesk verified from Annual Reports[27] | Organisation Internationale des Constructeurs d'Automobiles (OICA)[28] | Ecodesk. Verified from Carbon Disclosure Project (CDP)[29]

Table 4.3   Example of the Economic Benefits of Environmental Kaizen across Toyota (2008–2010)

No.	Description	FY 08	FY 09	**FY 10**
1	Reduction in energy costs	2.3	1.3	**1.5**
2	Reduction in waste processing costs	0.3	1.0	**0.3**
3	Sales of recyclable goods	12.4	4.4	**6.7**
4	Other (income from environment related technologies, etc.)	0.7	0.6	**6.8**
Total (Billion Yens)		15.7	7.3	**15.3**

recognized that economic growth and conservation of natural environment are compatible. Today, the Toyota Earth Charter has been adopted by more than 530 affiliates worldwide. More pragmatically, Environmental Committees, chaired by the company president, have built a tradition to investigate and develop responsive policies through strong collaboration among divisions.[30] For more than twenty years, the Toyota Guiding Principles and the Toyota Earth Charter have incrementally integrated the environment into company strategy deployment alongside the company's quest for superior quality and efficiency. In 2011, Toyota published a snapshot of the economic benefits of their environmental Kaizen. As illustrated in Table 4.3, *Toyota saved more than 38 billion yen or nearly a half-billion US dollars between 2008 and 2010 through continuous environmental improvement.*

Since 1994, environmental management plans have streamed more consistently into the annual plans of the different divisions across the world. Each year, environmental objectives are reiterated during the annual message of Toyota's president to reaffirm the company's commitment. Within Toyota, actions have matured from environmental risk audits and environmental assessment systems in the 1980s, to zero to landfill objectives in 1990s, to eco-buildings and eco-efficiency across production in the 2000s, and more recently, to transparent whole life-cycle thinking, which is unparalleled within the automotive sector.[31]

Indeed, given that the majority of emissions, or 75–85% of the end-to-end life cycle emissions of a passenger car, are during use,[32] it is important

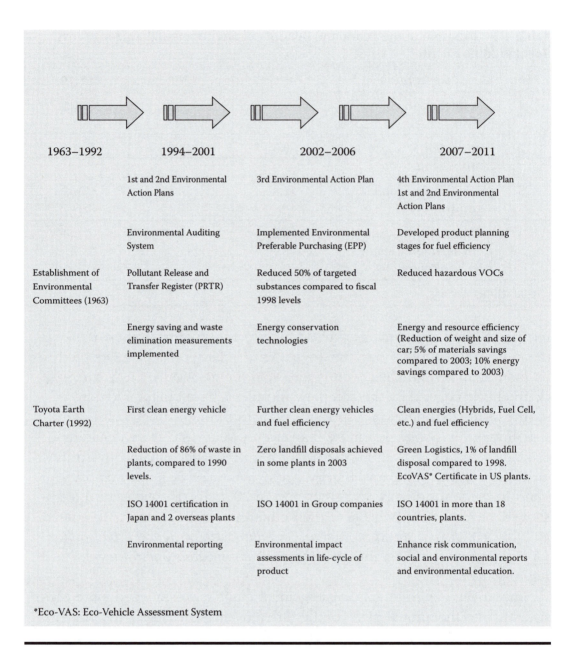

1963–1992	1994–2001	2002–2006	2007–2011
	1st and 2nd Environmental Action Plans	3rd Environmental Action Plan	4th Environmental Action Plan 1st and 2nd Environmental Action Plans
	Environmental Auditing System	Implemented Environmental Preferable Purchasing (EPP)	Developed product planning stages for fuel efficiency
Establishment of Environmental Committees (1963)	Pollutant Release and Transfer Register (PRTR)	Reduced 50% of targeted substances compared to fiscal 1998 levels	Reduced hazardous VOCs
	Energy saving and waste elimination measurements implemented	Energy conservation technologies	Energy and resource efficiency (Reduction of weight and size of car; 5% of materials savings compared to 2003; 10% energy savings compared to 2003)
Toyota Earth Charter (1992)	First clean energy vehicle	Further clean energy vehicles and fuel efficiency	Clean energies (Hybrids, Fuel Cell, etc.) and fuel efficiency
	Reduction of 86% of waste in plants, compared to 1990 levels.	Zero landfill disposals achieved in some plants in 2003	Green Logistics, 1% of landfill disposal compared to 1998. EcoVAS* Certificate in US plants.
	ISO 14001 certification in Japan and 2 overseas plants	ISO 14001 in Group companies	ISO 14001 in more than 18 countries, plants.
	Environmental reporting	Environmental impact assessments in life-cycle of product	Enhance risk communication, social and environmental reports and environmental education.

*Eco-VAS: Eco-Vehicle Assessment System

Figure 4.12 The evolution of Toyota's Environmental Achievements. Source: Various publicly available Toyota Environmental Reports.

to maintain a whole system perspective in understanding and reducing emissions within the automotive industry. For this reason, the Fifth Toyota Environmental Action Plan emphatically commits more to a wide range of actions to holistically reduce end-to-end environmental impacts. Toyota's key commitments in its Fifth Toyota Environmental Action Plan are to establish a low-carbon society, establish a recycling-based society, and protect the environment by creating a society in harmony with nature. Accordingly, Toyota has also developed a rather ambitious consumer-oriented *sustainable mobility strategy* that includes new-technology vehicles, innovative and bio materials (e.g., eco-plastics and water-borne paints), and more intensive diversification of energy sources to radically reduce dependence on fossil fuels. Toyota's sustainable mobility strategy for the next decades is filled with a combination of incremental and radical eco-innovations. At the heart of Toyota's sustainable mobility strategy is a *power train map* for the future of mobility, which incorporates a number of advanced technology platforms.[33]

For smaller sized vehicles and short-distance city commuters, Toyota has a vision of investing in electric vehicles (EVs). Toyota debuted the second-generation RAV4 EV, the RAV4 EV Prototype-Phase Zero, at an auto show in California in November 2010, targeting to achieve a 100-mile driving range. In 2012, Toyota also announced the imminent launch of the Scion iQ EV. The Scion iQ EV, which is a very small city car, has a range of less than 50 miles, making it suitable for urban fleets such as Zip Car.[34] While EVs have zero tailpipe emissions, Toyota expects only a small percentage of the market to be interested in EV technology due to drawbacks such as recharging time and limited driving range. In 2011, Toyota announced, "EVs will be one option in our portfolio of advanced technologies, but not the solution for every customer."[35]

As for mid-sized vehicles, such as the average size family sedans, Toyota has a vision of investing even further in its flagship hybrid and plug-in hybrid technologies. Currently, Toyota owns more than half of the hybrid car market and first signs of recovery as the first carmaker in the world are appearing.[36] In 1997, Toyota introduced Prius, the first-ever mass-produced hybrid power-train vehicle, to the global markets, and has already celebrated the sale of the one millionth Prius in the United States in April 2011. Toyota regards the hybrid technology as a stepping stone to minimizing the whole life-cycle environmental impacts of fuel-powered vehicles, and currently has nine different models within its hybrid range, excluding two new models of Prius planned for launch in 2012, one of which is going to be the plug-in hybrid Prius *c* Concept.[37] Toyota's plug-in hybrid offers the best of both worlds—the economy of an electric vehicle and the versatility of a hybrid fuel at the same time. Drivers have the choice to drive all electrically and recharge the on-board battery, or to benefit from the advantages and the utility of a conventional hybrid vehicle. It is estimated that it can reduce fuel consumption by up to one third over a conventional Prius.[38]

Finally, as for larger-sized passenger vehicles, vans, and even buses, Toyota has been investing heavily in Fuel Cell Hybrid Vehicle (FCHV) technology. In FCHVs, hydrogen is fed into a fuel cell stack where it is combined with air, creating a chemical reaction and generating electricity that in turn powers the vehicle's electric motor and charges the battery. FCHV exhaust emissions contain only water vapor and no particulate matter, GHGs, or other pollutants. Although the planned market launch of the first Toyota FCHV is no earlier than 2015, Toyota has already been road testing various advanced FCHV prototypes with various universities and government agencies in the United States, and achieving significant technological advancements while also reducing cost.[39] Toyota is also actively engaging with the hydrogen supply stakeholders to develop the much needed infrastructure for FCHVs.[40]

So far we have illustrated how Toyota is reducing its impacts on the environment through formation and deployment of a clear and consistent environmental strategy in line with its corporate Guiding Principles. We have benchmarked Toyota's performance against the rest of the automotive industry in terms of the total emissions from production and related areas, demonstrating that Toyota is consistently the best performer from various different perspectives. We have also showed how Toyota is heavily engaged in developing innovative solutions and technologies that will reduce or even eliminate the whole life-cycle emissions. It is evident that Toyota's sustainable mobility strategy, alongside its Five-Year Environmental Action Plan, paves the way toward a much more sustainable future, not just for Toyota, but also for the entire society.

Monozukuri: Harmony with Nature and Society

Toyota's fifth Five-Year Environmental Action Plan begins with the following statement of purpose: "Contributing to growth of sustainable society and earth through Monozukuri, co-existing with the global environment, making cars and offering quality products and services."[41] But what does *Monozukuri* mean? Monozukuri is a word that captures the true spirit of Toyota in relation to the concept of sustainability. The literal meaning of *Monozukuri* is "production." *Mono* is the thing which is made and *zukuri* means the act of making, but Monozukuri implies more than simply making things. It can be best compared to the word *craftsmanship* in English. Although in craftsmanship the emphasis is on the craftsman, whereas in

Monozukuri the person doing the making is deemphasized and the attention is on the thing being made.

This subtle difference reflects the Japanese sense of responsibility for using *things* in production and their deep respect for the world around them, both animate and inanimate. In the Japanese tradition of Monozukuri, the craftsman takes great care in using resources, not to be wasteful or futile. When an item or human effort is taken into use, there needs to be a benefit for the society as a result, while at the same time, the balance among production, resources, and the society should be maintained.

Monozukuri, therefore, is manufacturing that is in harmony with nature and that is value adding for the society. You could even say that *Monozukuri is the older sister of sustainable manufacturing.* Toyota's official website says, "Toyota has always sought to contribute to society through the Monozukuri philosophy—an all-encompassing approach to manufacturing. In its application of Monozukuri to the production of automobiles, Toyota has pursued a sustainable method of making its cars ever more safe, environmentally friendly, reliable and comfortable."[42] In 2008, Watanabe, president of Toyota Motor Company, delivered a presentation titled "Toyota for the Future,"[43] which illustrates how the ideas of Monozukuri and harmonious manufacturing are at the heart of what Toyota does.

Monozukuri is also about deeply respecting the individuals who do the job since there is no mindless repetition in Monozukuri. As in craftsmanship, in Monozukuri, workers "bring their mind to work" and are fully empowered and trained to deal with different situations, creating an elevated sense of ownership. Within Toyota, it is crucial for all workers not to be robbed of their right to pride of workmanship and to gain intrinsic satisfaction in what they do. In this concept, making products (Monozukuri) is also making people (Hitozukuri) because they are instilled with pride and passion for their jobs. Toyota's green vehicle technologies and other lean and green initiatives will not work without the full engagement of their people. Mutual trust, authority, empowerment, skills to make quality products, lifetime employment, and the inquisitive culture of *Genchi Genbutsu* (go see at the workplace) are all tenets through which Toyota respects its people.

In this way, Toyota mixes work with fun and the day job with passion. As a Toyota manager puts it,

> [I]n our North American Plants, teamwork and fun play a very important role. We have Kaizens through car or horse tracks … which are visible among all work teams and then it becomes a

competition to report monthly your amount of reduction of energy and reduction of costs. Team members take home what they learn and worry about energy reduction and recycling at their homes as well. We have as well treasure hunts. They go to different sections of the plant or different plants and then it becomes a competition. I have sent our engineers on the weekends to look at ways to reduce energy in communities, like Newport Aquarium. We do this for plants, our suppliers or communities."[44]

Conclusions

Overall, Toyota's environmental initiatives are driven far more by their profound commitment to harmonious manufacturing (Monozukuri) and their role in the society as a value-adding corporate citizen dating back to the precepts of the founding father Sakichi Toyoda, than by environmental regulations. We have demonstrated that they perform incredibly well against the rest of the industry and that they have rigorous structures for developing environmental strategies and integrating them into their daily operations through meticulous Hoshin planning. Moreover, we have shown how Toyota has developed a holistic view of the future focusing on short-, medium-, and long-term technological and managerial solutions to reduce its environmental footprint holistically. Toyota's remarkable success in hybrid technologies is soon to be followed with the introduction of several plug-in hybrid, electric vehicle, and fuel cell hybrid vehicle platforms.

The resilience of the Toyota Way was tested once more during Toyota's recent crisis. Many commentators and industry experts have said that Toyota weathered the storm well[45] and has used the problems as an opportunity to challenge itself toward a more sustainable future. Toyota did this not by doing something different, but by going back to its basics and by an even greater commitment to the Toyota Way. Throughout history, challenges, large and small, have reinforced the Toyota Monozukuri discipline and stoked its passion for innovation. As Toyota looks toward 2020, many new challenges and opportunities await. By going back to its guiding principles and by practicing the idea of harmonious manufacturing, Toyota has every chance to shape the future of the automotive sector and maybe even the future of all manufacturing.

Endnotes

1. For further information, we suggest reading "The Seven Value Stream Mapping Tools," by Peter Hines and Nick Rich, *International Journal of Operations and Production Management* 17, no. 1 (1997).

2. The content of this case study is the authors' own opinion and do not necessarily reflect that of the Toyota Motor Company.

3. *Toyota Environmental and Social Report*, 2003. Toyota Environmental and Social Report, p. 4, (online) http://www.toyota.co.uk/br/aboutus/E_kankyouohou-koku2003.pdf.

4. A. Mass and A. Robertson, "Organizational Innovation in the Toyoda Enterprises, 1895–1933," *Business and Economic History* 25, no. 2 (1996): 4.

5. H. Satoshi, *Inside the Mind of Toyota: Management Principles for Enduring Growth* (New York: Productivity Press, 2006), 15–23.

6. H. Shiomi and K. Wada, *Fordism Transformed: The Development of Production Methods in the Automobile Industry* (Oxford: Oxford University Press, 1995).

7. A. Mass and A. Robertson, "Organizational Innovation in the Toyoda Enterprises," 35.

8. M. Holweg, "The Genealogy of Lean Production," *Journal of Operations Management* 25 (2007): 421–423.

9. T. Ohno, *The Toyota Production System: Beyond Large-Scale Production*, (Cambridge, MA: Productivity Press, 1988), 75.

10. Toyota Motor Corporation, "Toyota Production System," Toyota website, 2012, http://www.toyota-global.com/company/vision_philosophy/toyota_production_system/

11. T. Ohno, *The Toyota Production System: Beyond Large-Scale Production*, (Cambridge, MA: Productivity Press, 1988), 75.

12. Ibid.

13. Toyota Motor Corporation, "Human Resources Development," 2001 Environmental Report, http://www.toyota.co.jp/en/environmental_rep/03/pdf/E_p80.pdf.

14. J. P. Womack, D. T. Jones, and D. Roos, *The Machine that Changed the World*, (New York: Simon and Schuster, 1990).

15. Ibid.

16. Josh Cable, "GM, Ford and Chrysler Strive to Become the Lean Three," *Industry Week*, November 18, 2009, http://industryweek.com/articles/gm_ford_and_chrysler_strive_to_become_the_lean_three_20441.aspx?Page=3?ShowAll=1.

17. N. Stern, *Stern Review on the Economics of Climate Change*. HM Treasury, London Executive Summary, 2006, http://www.webcitation.org/5nCeyEYJr.

18. John DeCicco and Freda Fung, *Global Warming on the Road*, Environmental Defense, 2006, http://www.edf.org/sites/default/files/5301_Globalwarmingontheroad_0.pdf.

19. Ibid.

20. Hyung Chul Kim, "Shaping Sustainable Vehicle Fleet Conversion Policies Based on Life Cycle Optimization and Risk Analysis," PhD diss., University of Michigan, 2003.

21. In our benchmark, total emissions include exhaust from company facilities, exhaust from transportation, company vehicles and machinery, and indirect emissions related to purchased energy (gas, electric, etc.), but excludes indirect emissions related to business travel, employee commuting, and a few areas of transportation. In terms of the accounting methods used by the Carbon Disclosure Project (CDP), we have used Scopes 1 and 2 and not used Scope 3. The excluded items are reported to be a small percentage of the total. However, only a few companies (such as Toyota) make Scope 3 data available based on the latest CDP information, and we have excluded for the sake of consistency.

22. Toyota Environmental Report, 2011.

23. These are the words of Professor Kaoru Ishikawa, father of Hoshin planning, in his book: *What Is Total Quality Control?: The Japanese Way* (Prentice Hall: New Jersey, 1985).

24. Information compiled from various Carbon Disclosure Project (CDP) annual reports and individual company profile from Ecodesk.

25. Toyota Motor Company (2010), Toyota Global Vision, [online] www.Toyota-global.com

26. www.ecodesk.com — various company profiles.

27. Ibid.

28. www.OICA.net

29. www.cdpproject.net

30. *Toyota Environmental Report*, 2011. Toyota Motor Corporation (2011) Toyota Environmental Report, [online] www.toyota.no/images/environmenal_report11_fe_tcm308-1100813.pdf

31. Lounsbury, E. (2011) "International Carbon Flows", Low Carbon Vehicle Partnership and IMechE Life-cycle Seminar, Nov. 2011. http://www.google.co.uk/url?sa=t&rct=j&q=&esrc=s&source=web&cd=1&ved=0CDcQFjAA&url=http%3A%2F%2Fwww.lowcvp.org.uk%2Fassets%2Fpresentations%2FEric%2520Lounsbury%2520-%2520Carbon%2520Trust_ICF_LowCVP-IMechE%2520Life-Cycle%2520Seminar.pdf&ei=CZDPUN6mPPSW0QWhw4DwBw&usg=AFQjCNEOt7rIeRb0gMbY7eipmnnkyXHqfA&cad=rja

32. G. Smith, "Toyota's Environmental Technologies and Approach," Presentation at Cambridge University Energy Network (CUEN) Annual Energy Conference on Sustainable Transport, 2010. http://www.cuen.org.uk/imag/Academic_material/4%20Gram%20Smith.pdf

33. Peter Valdez-Dapena, "Cars of the Future: They're Going to be Tiny and Weird—Scion iQ EV," CNN Money, April 12, 2012, http://money.cnn.com/galleries/2012/autos/1201/gallery.future-city-cars/4.html

34. *Toyota North America* (2011) *Toyota North America Environmental Report 2011, Challenge, Commitment, and Progress*, p.20.

35. According to www.edmunds.com from 2002–2011.

36. *Toyota North America Environmental Report*, 2011.

37. *Toyota North America Environmental Report*, 2011.

38. *Toyota North America Environmental Report*, 2011.

39. G. Smith, "Toyota's Environmental Technologies and Approach," Presentation at Cambridge University Energy Networks Annual Conference on Sustainable Transport, 2010.

40. Toyota Motor Corporation, "The Fifth Environmental Action Plan," Toyota news release, 2011, http://www2.toyota.co.jp/en/news/10/08/0825b.html.

41. Toyota Motor Corporation, "Special Story: Contributing to Society by Making Cars - Toyota Contributes to Growth of India's Economy and Society Through Etios," 2011, http://www.toyota-global.com/sustainability/sustainability_report/special_report/contributing_to_society_by_making_cars.html.

42. Katsuaki Watanabe, 2008, "Toyota for the Future: In Pursuit of 3 Sustainabilities," June 27, 2008.

43. Kevin Butt, (2010) Speech delivered at From Shop Floor to Top Floor: Best Business Practices in Energy Efficiency Conference, Centre for Climate and Energy Solutions, 2010, accessed January 22, 2012, http://www.c2es.org/energy-efficiency/conference/speakers/kevin-butt-general-managerchief-environmentalsafety-officer-to

44. J. K. Liker and T. Ogden, *Toyota under Fire: Lessons for Turning Crisis into Opportunity* (New York: McGraw-Hill Professional, 2011).

Chapter 5

Lean and Green Business Leadership

In this chapter we focus on the leadership and people engagement element of the Lean and Green Business System model (Figure 5.1). This chapter tells the story of Adnams brewery's journey as an exemplary case study of lean and green leadership. This is not only a readable story, but one that has respect for humans at the heart of it.

We highlight and explain the key features of lean and green leadership through the lens of this case study. The case study shows how giving direction and focus to the organization is everyone's job and not just the job of senior managers. The case study also highlights the significance of committing to a set of sustainable values, engaging everyone in change toward those values, and pursuing them with persistence at all times. Moreover, for managers who want to embark upon their own lean and green journey, this is an excellent example of why good leadership lies in understanding the work, the variability of the work, and the system, as much as it is about dealing with human behaviors. In the previous chapters we discussed how deep appreciation of the system depends on frontline understanding of work. This is what Toyota calls Genchi Genbutsu (go to workplace and see) and what Adnams has put into practice.

For senior executives and middle managers alike, the best psychology for managing change and leading people is to be genuine about one's commitment to lean and green. In the case of Adnams, team members are lending their support willingly because they see top management's true commitment to their vision of building a more sustainable company, in action and beyond words.

Figure 5.1 Lean and Green Business System model.

Adnams strongly reflects the central role that committed, inspirational leadership can play in making fundamental changes to business operation, while at the same time facilitating a genuinely bottom-up approach, and thereby engendering the widespread dedication to sustainable business that so many businesses appear to lack. The case study also highlights how innovation and new technologies can be harnessed to simultaneously reduce environmental impact, improve operational efficiency, and increase profitability. But more importantly, it provides deep insights into the leadership qualities that drive innovation to precipitate lean and green results.

Lean and Green Leadership: Case Study of Adnams PLC.

Introduction

On May 1, 2008, Adnams Brewery, based in Southwold in Suffolk, launched a line of bottled beer called East Green. For the keen beer drinker, the beverage is a light golden beer with subtle citrus and grassy hop aromas, dry and refreshing, with well-balanced bitterness. However, more importantly, *as Europe's first-ever carbon-neutral bottle of beer*, the product line stands as the celebratory culmination of an outstandingly unique story of lean and green business practice. The example of Adnams brewery is an exemplarily case study of the leadership's profound commitment to sustainability and a down-to-earth realization that reducing environmental impacts is essential to economic success. As Adnams' CEO explains: "We see the rise and fall of the North Sea every day. When the high water line coincides with the spring tide and a 'North Sea surge,' some of the beach huts on Southwold sea front get washed away, some of our pubs get flooded, and we can't get out lorries out of the distribution centre. We live with the elements; we are on the edge of the elements and we need to take account of that in our business planning."

In this example, we discuss the lean and green innovation story at Adnams, while paying special attention to the people engagement and leadership issues by highlighting specifics about the backgrounds and characteristics of some of the key individuals involved in the Adnams journey.[1]

Figure 5.2   Adnams eco-efficient distribution center in Southwold, United Kingdom.

Adnams Brewery: A History of Innovation and Community Values

The first recorded instance of brewing activity taking place on the current Adnams site dates back to 1345, when Johanna de Corby and the 17 "ale wives" of Southwold in eastern England were charged by the manorial court with breaking the assize of ale. Beyond this, little is known up until George and Ernest brought the Sole Bay brewery with the help of their father in 1872 and the name of Adnams first became associated with the brewery. After the departure of George (for a life of adventure in South Africa, where he was unfortunately eaten by a crocodile), Ernest Adnams and his head brewer, Thomas Sargaeant, signed the Memorandum of Association establishing Adnams & Company Limited in 1890; in 1902 Jack and Pierse Loftus acquired a stake in the brewery. The brewing enterprise developed well after the turn of the century and by the 1950s, Adnams had begun to earn a worldwide reputation, for example, winning the Star of Excellence in Brussels in 1953.

It was not, however, until the 1970s that the bulk of Adnams' expansion took place. The brewery was modernized and new systems introduced to cope with the greater demand. Exports gradually traveled further afield and Adnams beer appeared increasingly in pubs and bars across the United Kingdom. Today, the brewery produces mainly cask ale and pasteurized bottled beers, although until 2006 it maintained the tradition of delivering locally with a horse and cart. Currently, the output of the organization is around 85,000 barrels of beer, including bottled beer, which are distributed mainly through the East Anglia region and in London via some 1,000 outlets countrywide. The current turnover of Adnams is more than £50 million or US$75 million and the company now employs some 420 people.

The growth of Adnams has been facilitated partly by finding new ways to do what the company has always done well; for example, through the 1999 relaunch of its flagship beverage, Broadside, in award-winning bespoke 500-ml bottles. However, Adnams has also been a pioneer of innovation despite working within a largely traditional sector. For example, the brewery offers a "Beers of the World range," which includes Kolsch, a golden ale based on a warm-fermented style from Cologne in Germany; a Belgian Abbey beer; a German wheat beer; an American style IPA; and an Irish dry stout. It is also significant that Adnams sells something in the region of 93% of its beer outside its own estate and even into a widening international market, in which Adnams can be found in the Anglaise sections of Southern French supermarkets, for example.

The second string to Adnams' innovation has been diversification into new areas of business. As might be expected, the company now owns over 70 pubs, but also 2 bespoke hotels that offer standard accommodations as well as seasonal and special stays, such as a Christmas shopping break. The restaurants on the Adnams estates also go a little further than most, having attracted head chefs of the highest caliber. Even less traditionally, Adnams has displayed great entrepreneurial thinking through the establishment of a wine importer, both for use on the Adnams estates and to be sold as a retail interest through the Adnams' Cellar & Kitchen stores, which specialize in high-quality kitchenware ideal for those taking a Christmas shopping break in Southwold town. The first of these outlets opened in Southwold, but the chain has quickly multiplied since 2005, and now consists of some 10 stores across East Anglia, London, and Lincolnshire.

On the production side, Adnams commissioned the Copper House Distillery on the site of a disused brewhouse at their Southwold site in November 2010. While the new development faced inevitable technical challenges in the early days of its operation, learning has been rapid and the enterprise now represents a fine example of a craft distillery, generating a range of products that have already won 10 international awards, including a Gold Medal at the San Francisco World Spirits Competition. Naturally, this venture also allows further creativity in other areas of the business, for example, the opportunity to offer spirit tasting menus in the company's hotels. To place these initiatives in context, a senior representative explains that while revenues from new product development used to be only 4%, in 2011, 22% of revenues were coming from new product lines.

What is truly special about the case of Adnams brewery, however, is the innovation in sustainability practice that has permeated all aspects of their business operation. Indeed, while the company has a longstanding history of responsibility, the official and genuine embedding of strong social and environmental values at the heart of the company has been a key feature of the *Adnams way*. These values, combined with technological excellence, have not only enabled substantial improvements in terms of environmental performance, but have generated quantitative and qualitative business benefits principally in the form of cost savings and brand development.

For many years, Adnams brewery has had a strong sense of its important role within the local community. Put another way, *responsibility* has been a fundamental part of the company's DNA. The company has played a large part in the regional economy as the biggest local employer and has focused its purchasing on suppliers close to home. This concentration of activity includes the relations established over the years with something in the region of 40 local East Anglian farmers who supply the grain for malting, but also growers of strawberries and butchers of meat used in Adnams pubs and restaurants. As the operations director describes, "There's not a business this size for something in the region of 12 miles. We like to buy local produce; we like to employ local people. It just bodes well for the area as, in honesty, there is not a lot else here."

According to internal stakeholders, the somewhat unintentional lean and green approach to business has picked up this legacy of community responsibility and run with it. Reportedly, the process began with the arrival of Dr. Andy Wood, who is currently the company's CEO (see Box 3). Operations officer Karen Hester (who was working in procurement at the time, *see* Box 2), recalls Andy often speaking about the issue of rising energy and utilities costs, and how this would have a significant impact on businesses if they were not factored in planning.

When Andy Wood moved on to the board of directors in 2000, he drew on this philosophy in introducing a set of company values (see Box 1) which, according to the management, "we want employees to live and breathe, rather than just pay lip service to." Indeed, in Andy's view, "We don't actually talk about Corporate Responsibility or Corporate Social Responsibility. My view is that if you have to use acronyms, or if you have to produce a separate Corporate Social Responsibility report, you actually don't really get it. What you should be looking to do is to fundamentally embed these things in your business so they influence every part of your operation."

While Adnams is an incorporated PLC, and therefore financial issues remain central to the decision-making process, the management note that "we also have another filter in our decision making. It gives us the opportunity to ask if a certain possibility lives up to our values."

Principal among the Adnams company values is the necessity of working toward enriching the lives of both customers and employees of the brewery. However, as part of this, it is also felt that impact on the social, natural, and built environment should be rendered as positive as possible.

Setting out company values has been integral to the approach of most companies to develop a corporate social responsibility (CSR) strategy. However, Karen Hester, the operations director, says that the first real environmental initiative was to communicate the genuine importance of these values to the workforce in a way that they understood, because "you always

BOX 1: INSPIRING CHANGE FROM THE BOTTOM-UP: THE ROLE OF VALUE-BASED MANAGEMENT

ADNAMS' COMPANY VALUES

Fulfilment

We want fulfilled customers and employees, whose lives are enriched by their involvement with Adnams.

Quality

We want to create the best products and offer the best service, and are always looking for ways to improve.

Environment

We aim to manage our impact positively on the social, natural, and built environment.

Integrity

We deal with people openly and honestly, building strong, supportive relationships.

Commitment

We expect commitment to these values, and aim to translate them into everyday realities.

have to adjust your communication to your audience," and therefore generate genuine and fundamental staff buy-in. Put another way, and no doubt echoing the brewery dray drivers from the past, "you can lead a horse to water, but you can't make it drink," Karen says. Indeed, before the launch of environmental values, there were many issues in the company, for example, "people would leave all the lights on, or all the heating on, and go home because it was not their problem; there were loads of little things."

One early initiative was the inclusion of environmental incentives within the scheme of profit-related pay. At the time, this included a 4% bonus for key performance indicators (KPIs), 1% of which was directly related to a departmental environmental initiative. Given the variation in environmental impact, each department was given autonomy to set its own projects and targets (comparing impact from the same quarter one year to the next), as

BOX 2: KAREN HESTER, CHIEF OPERATIONS OFFICER

THE LEAN CONTRIBUTION TO OUTSTANDING PERSONAL DEVELOPMENT

Originally from the Southwold area, Karen joined the Royal Logistics Corps of the British Army at the age of sixteen. Karen became the youngest female to obtain her Heavy Goods Vehicle (HGV) license and gain her first stripe in record time. The fact that she rejected transfer to office training in favor of working her way through the noncommissioned ranks testifies to her perspective on hard work.

After being dismissed from service for becoming pregnant, Karen signaled a willingness to facilitate fundamental change. With others she fought and won a landmark legal battle in which the rights of military women were drastically changed.

Moving back to Wangford, Karen joined Adnams brewery as a part-time cleaner in 1988. Moving to a full-time job in procurement in 1990, Karen remembers the arrival of Andy Wood as the new logistics manager, partly as his would be the first car she ever bought.

Every Friday, Karen would fill Andy's car with petrol and "chit chat." She remembers remarking that he stood out, having different ways of looking at things and thinking about things, and she predicted to her then boss that he would be the first person from outside the community to join the board of directors. One of these differences was his vocal and early concern with the rising cost of energy and utilities and clear enthusiasm that these factors should be part of business planning.

After Andy learned about Karen's past in the military, he invited her to move into the role of transport clerk, from where she soon moved to act as transport manager. At this point Karen obtained her civilian Certificate in Professional Competence, Certificate of Management Studies, and Diploma in Management Studies, studying evenings supported by Adnams.

As her dissertation for the latter qualification, Karen evaluated the proposal to move Adnams' distribution center to newly acquired land in Raydon. The project was given the go-ahead and began in 2005, at which point Karen moved to project manage its development.

Since then Karen has moved through managerial roles, becoming operations director in 2007 with wide responsibility for the brewery, human resources, information technology, pubs, and hotels.

As a testimony to her career, Karen was awarded East of England Business Woman of the Year 2008. However, what is perhaps a more pertinent insight is that she immediately wrote personal letters to all of her team to thank them for their contributions.

BOX 3: DR. ANDY WOOD, CHIEF EXECUTIVE OFFICER

Like many people who go on to successful entrepreneurial careers, Andy left school at the age of sixteen, and went to work for a large financial services company. There, Andy received on-the-job training and development, and learned an incredible amount—including, he says, many new ways of looking at things. Having caught up quickly with the graduate intake, he spent fifteen years in the company after which he was supported to complete his MBA.

After this training, Andy says that like many people who go through the same experience, he decided that he wanted to move on to a new challenge. More specifically, Andy's last role in the large organization involved the promotion of a values-led approach to business operation. While he remains tremendously proud of the achievements that his former company made in that area, he still remembers being frustrated by the limitations. "No matter how many little fires we lit under that organization, the weight of corporate water fell on it and put those fires out, and I think that was an opportunity missed," he explains.

With this experience under his belt, Andy wanted to move on from a large organization to a somewhere smaller one in which he felt he could make more of a difference. Although he is not shy to admit that this "sounds a bit cliché," he says honestly that "this was my thinking."

Andy joined Adnams in 1994, with the initial remit of carrying out reforms in the supply chains and improving the customer service ethos. At the time he remembered that he expected that he might stay for 3 to 5 years; however, things have worked out very differently.

Right away, Andy recalls, he was made to feel welcome as one of few outsiders at the time. "I was really struck with the management team here and the opportunities that there were." He recalls, "They let me get on with it and that was fantastic."

As Andy's work was appreciated by the company's directors, he was moved through various jobs with increasing levels of responsibility. However, as he has moved up the company, he has certainly not forgotten his passion for values-led operation.

Andy says, "It has become very important to me, this whole notion that companies have a purpose that is beyond just delivering shareholder value at all cost … this whole approach to values, and what an organization stands for has been critically important both to my own personal development but also the development of Adnams I think."

Andy is, however, clear to point out that the existing ethos at Adnams allowed him to put his ideas into practice. "Family ownership encourages a long-term perspective" he thinks, and the current board of directors, led by Jonathan Adnams (who is clear that "We have to give something back. Where is the business case for contributing to the destruction of the planet"), sees the importance of this view; for example, William Kendal is an organic farmer and Steve Sharp has worked on Marks & Spencer's Plan A in his role as their marketing director.

The other notable element of Andy's perspective is that he does not accept CSR as an adjunct of business—something to be run by the marketing department, or Corporate Affairs office, and written up in a separate report. Furthermore, this is a very strongly grounded intellectual position, as he references the work of Michael Porter, the Harvard professor, as wanting to move social responsibility beyond a defensive box ticking exercise, to an offensive and positive competitive corporate strategy.

"That is what we are trying to do here," says Andy. "At the heart of what we do is a set of behaviors, our values, our organizational code."

Perhaps one of the things that sets Andy apart from most business leaders is that he has a personal interest in climate change. He believes "that climate change is happening, and that man has contributed to this. I think that companies have a leadership role in changing the way in which the economy operates. I have a huge sense of pride for what Adnams does and what its potential is."

long as these were approved by senior management. However, it is emphasized that while this initiative was about embedding a sustainability ethos in the workforce, it was also about fostering a bottom-up approach. Indeed, the operations director at Adnams remarks that "the best ideas have come from the shop floor." To maintain this engaged approach to work, Adnams now pays 8% performance-related pay (PRP), within which reducing environmental impact has remained part of staff evaluation.

As well as this strategy, some arms of the company still have environmental champions. These personnel have a responsibility to ensure good environmental practice as part of their jobs, and are particularly important in parts of the business where technological aids cannot be built into the fabric of the business infrastructure, for example, in a historic building where planning law restricts the installation of motion sensors to turn off lights.

All in all, the embedding of environmental responsibility has been highly successful. In the 2011 staff survey, almost all staff members (bar three or four out of the total) said the company values were important to them. This is obvious in the day-to-day practices, even to the untrained observer; for example, while dining in the restaurant, the manager can be seen making a considered decision if lights are needed at that time in the afternoon.

This environmental awareness is also fostered by regularly allowing Adnams staff the space to interact with decision makers and highlight their observations and propose innovations of their own. The management has an open-door policy, which is more than evident in the ratio of conversations held to meters walked by key personnel as they move around the Adnams site, and even across the town as a whole. Management also runs "snack and chats" on a regular basis. Here the company provides a buffet lunch and gives the workforce space to discuss any issues with the senior management. As a result, in the most recent staff survey, which again aims to reduce hierarchical relationships by inviting the workforce to provide anonymous feedback on named members of the company's management, close to 100% of the workforce feel that the senior management are approachable and responsive to employee views, and a wide number of respondents took the time to make free comments on the company's family feeling and approachable nature of managers who care about community values.

Herein lies an important similarity between Adnams and Toyota. Both have a strong family business spirit and both take the principle of Genchi Genbutsu to heart. As discussed in the previous chapter, Genchi Genbutsu means *going to workplace* for understating and being proximate to *the way the work works*. Moreover, both Adnams and Toyota are value-driven organizations. They take lots of time and meticulous care to articulate their vision and even to codify these values if required.

While this space has provided employees with the opportunity to highlight issues, such as problems with the car park and requests for high-visibility jackets, environmental initiatives have also been proposed. Indeed, management at Adnams says they are "most proud that we get little ideas bubbling up from the staff … where they can make a real difference." For example, one staff member identified a much less environmentally impactful and less expensive way of keeping clients warm in the colder weather. In place of providing gas heaters in these external spaces, which is a widely favored strategy for most pub chains, Adnams staff located waste textile materials, had them stitched into rugs and adorned with the Adnams logo, so that they could be distributed to clients to keep them warm. Not only did this reduce the use of gas, save money, and find a use for otherwise redundant textiles, but also allowed Adnams to demonstrate in practice their view of customer services: "When our customers get cold in the evening now our staff go out and wrap them in a blanket. That is more a metaphor for the way that we like to look after our customers," Karen Hester says.

Where Values Meet Innovation: Sustainability Investments at Adnams

Energy-Efficient Brewstream

In response to increasing demand, Adnams set about renovating its production systems by installing new state-of-the-art fermenting vessels in 2001. However, in 2006–2007, the company completely replaced its 100-year-old brewing system, which was probably the first new brew system installed in the UK in thirty years. In line with the firm's focus on sustainable development, Adnams opted for the most efficient systems available on the market.

The new brewing kit was modeled on a design from mainland Europe that incorporates a mash conversion vessel, a filtration vessel, and a whirlpool kettle to boil the hops. The kettle is designed so that the heat generated during the boil, and previously released into the atmosphere, can be reused for the following brew. In this way, the technology recycles 100% of the heat used in order to provide 90% of the energy for the next brew. "The process of brewing is quite energy intensive," says Andy Wood, and "boiling water by definition is an energy-intensive business. And that's why we've tackled this."

Figure 5.3   Adnams state-of-the-art brewing facility saves an average of 30% on gas and water each year.

In addition to this, there is a 97% extract rate from malt and hops, which is a much higher level than in conventional breweries. Fermentation takes place in conventional square vessels that are enclosed to prevent the escape of carbon dioxide. All vessels are filled from below to stop oxygen pick up. The beers need 2 days of primary fermentation followed by 7 days conditioning. The new plant has the only wet mill in the country that cuts oxygen when malt is crushed in preparation for brewing. On the whole, this equipment saves an average of 30% a year on gas and water consumption while increasing flexibility to brew different styles, using malts and hops from other countries (Figure 5.3).

The efficiency of the brewing equipment is also aided by the complete computerization of the processes, and this significantly reduces human activities and flows of material inside the brewery, minimizing potential areas of waste inefficiencies.

As a result of these innovations, while the old system required 8 pints of water to make a single pint of beer, the new system uses around 3.1 pints; and furthermore, with continual improvement through adjustment of the evaporation levels, usage is now down below 3 pints of water for every pint of beer produced.

While the good financial sense of reducing water consumption has led many breweries to also reduce their usage, the opportunity to replace the entire brewstream has allowed Adnams to go much further than most.

While some of the efficient operations still use something in the region of 4.5 pints of water to produce a single pint of beer, the industry average is still 6 and 8 pints. The Adnams brewstream thus represents a substantial movement forward in terms of eco-efficiency and cost reduction; although, as might be expected from the embedding of sustainability values; the objective of the company is to further reduce the quantity of water used.

Eco-Efficient Distribution Center

As demand for Adnams beer continued to grow, the company began to struggle to service this from the small site within the confines of Southwold town center. Therefore, in 2006, Adnams decided to build a separate distribution center in line with the following three requirements:

- Minimal environmental impact
- Maximum operational efficiency
- Superior returns on investment

To begin with, the selected site in the neighboring village of Reydon was that of an old gravel pit and, hence, involved the redevelopment of an old brown field site. Immediately after purchase of the 88 acres of grassland in 1996, Adnams developed a nature conservation scheme in consultation with the Suffolk Wildlife Trust: replanting hedges and a new wood, and dredging and contouring existing ponds to encourage wildlife.

Despite these efforts, the proximate location of the site to the brewery offered to maximize the efficiency of the production process. Also, the relocation of Adnams Warehousing & Distribution, Cellar Services and Customer Services (which previously operated from an increasingly cramped site close to the heart of Southwold) offered to significantly reduce journey miles for Adnams employees and delivery vehicles coming in from the only entry to Southwold on the east. Indeed, it is estimated that there has been a reduction of approximately 60 heavy goods vehicles and 20 van deliveries a day in and out of the Southwold, as well as a reduction of some 60 employee cars, entering, leaving, and parking in the town, all of which has reduced noise and disturbance for local residents.

As the surrounding area is classified as one of outstanding natural beauty, Adnams had a clear aim to minimize the visual and light pollution. In order to extend their lean and green operation, Andy Wood suggested that as well as ensuring that local architects had the opportunity to tender for the opportunity to design the new distribution center, Adnams would also make an effort to cast the net wider in order to get some of the latest architectural ideas (a perspective that demonstrates a reflexive view of sustainability as opposed to assuming local is best).

The winning architect, London-based Aukett Fitzroy Robinson, provided the company with a "green shopping list," laying out possible sustainable features and benefits, as well as any financial payback times, from which the board could then choose. One of the big desires was that the storage area, which for the quality storage of ale needs to be kept at 13°C, should not need to be artificially cooled or heated.

Adnams also enacted its principle of promoting contributions from the tacit knowledge of its workforce. For example, having worked in warehousing and logistics, Karen Hester was able to make the suggestion that the distribution center would benefit from not having supporting columns for the roof. The reason for this was that supporting structures on the warehouse floor reduced the versatility of the warehouse space, as deliveries of different sizes constantly had to be rearranged to fit. This created inefficiencies and reduced the welfare of employees required to physically move such items which, even with tight health and safety procedures, generated fatigue and injury problems.

As a result of this process, the structural heart of the new distribution center features a supporting steel frame with Lime Technology's *Sumatec unfi* blocks, each weighing about 19 kg and made to size at 100 × 254 × 356 mm. These blocks are produced by compressing a mixture of red lime and earth and forming a diaphragm wall. However, also fundamental to the design is what, at the time of construction, were the largest single span (90 m from end to end) *glulam*, or wood laminated beams, in the United Kingdom. These reduce the environmental costs associated with steel (embodying 24 times less energy per ton of finished product) and have been sourced from sustainable forestry in Denmark. As the plan required the building to be invisible from the road, the structure was set seven meters underground, and the previous use of the site as a gravel pit negated the need for significant excavation. Furthermore, the roof is constructed from sedum, a substitute for grass. This not only merges with the natural environment but helps to regulate the internal temperature. Rainwater is captured by the building, filtered through adjacent reed beds, and returned to the building for reuse for vehicle washing, staff showers, and flushing loos.

In order to obtain a suitable material for the rest of the building, Adnams worked with Bath University, investing some £60,000 to £100,000 to revive, develop, and market a building material used many years ago in continental Europe. For the architect, hemcrete, a concrete-like substitute made from hemp and lime, and which is now marketed by Lime Technology, was the solution to creating an ultrasustainable, eco-efficient building. Fundamentally, hemcrete is a mix of lime and hemp that offers instant environmental savings because the materials can be locally sourced. In this case the lime (a waste product from quarrying) from Derbyshire and the hemp from East Anglia itself. The 90,000 blocks required could be manufactured using less energy than required to produce conventional concrete components.

In between this internal and external block-work, Tradical® Hemcrete was applied as an infill providing the building's necessary insulation. *The resulting high insulating properties of the whole structure enabled the 4400-m² Adnams distribution center to naturally maintain an internal temperature at between 11 and 13°C.* All entrances to the warehouse incorporate buffer space with an internal and external set of doors that are timed to allow vehicle emissions to dissipate so as to preserve the internal temperature. By obviating the need for a mechanical cooling or heating system, this innovative element fulfills the vital criteria of the building, which is used to store thousands of bottles of beer and wine.

Finally, *Limetec*, Lime Technology's own brand of hydraulic lime mortar, plaster, and render, was used to lay the hemp blocks and permit long-term durability and weather protection of the walls. Lime/hemp construction generates considerable environmental benefits, notably in terms of thermal performance, regulation of humidity and temperature, as well as carbon sequestration. The technology allows for outstanding thermal performance in that, while standard U Value regulations are set at 0.35, the building achieves 0.18 in addition to providing excellent sound insulation. The passive regulation of humidity and temperature means that no mechanical heating or cooling systems are needed in the warehouse. Likewise, the office space above the center is kept cool by natural ventilation and roof overhang. Also, while a conventional brick and block building of a similar size would generate about 300 to 600 tons of CO_2 emissions, Adnams' eco-efficient distribution center locks up 100 to 150 tons of CO_2 within the walls, a saving of at least 450 tons of CO_2 by using lime/hemp construction. This is because in common with all similar plants, hemp captures CO_2 and releases oxygen during its rapid growth. Furthermore, the lime mortar and render, used to face the lime hemp blocks and provide long-term durability and weather protection, actually absorbs carbon dioxide as it sets. The immediate and positive effect of this process is the sequestration of the principal

Figure 5.4   Adnams eco-efficient distribution center was constructed with natural material.

greenhouse gas. Each block produces one tenth of the CO_2 emissions of a conventional concrete block. Even more carbon dioxide absorption occurs as the air-lime-based binder in the product sets.

Andy says that the decision to build this eco-efficient distribution center was driven equally by environmental credentials and traditional business sense. Indeed, while there is a heating system for use only in the coldest weather, the innovative construction enables Adnams to save 50% on electricity and gas and, therefore, constitutes a substantial cost savings (Figure 5.4).

As with many other environmentally friendly concepts, a green warehouse triggers a string of further positive effects in the building supply chain. For example, the use of hemp includes even more benefits, such as promoting the production of an environmentally sound commercial crop as the plant does not need chemical spraying because it suppresses weeds and pests. When mixed with Tradical to make walls, it provides high thermal and acoustic insulation, excellent water vapor permeability, high flexural strength, and high carbon capture.

As part of the project's further green credentials, a number of additional features can be highlighted:

■ The building features a green sedum roof that filters pollutants from the air and helps purify it, while providing a natural habitat for fauna that is lost in standard construction processes.

■ The wooden glulam roof support beams provide a column-free interior and roof lights.

■ Car parking areas are made of Golpla®, a strong, honeycomb mesh made from low-density recycled plastic, which controls erosion and storm water, provides sustainable drainage, reduces concrete use, and blends in with the surrounding grassland.

■ Clerestory glazing provides natural light to the interior.

■ Solar collectors mounted on the roof provide 80% of the site's hot water.

■ Movement sensors installed throughout the warehouse stop electricity being wasted.

■ Rainwater harvesting and a foul water waste system, including septic tank and reed beds, enables water to be cleaned and delivered to adjacent ponds.

The building project clearly manifests the commitment of Adnams to innovate sustainably. In fact, with so many sustainable elements, the building, which now acts as Warehousing and Distribution, Customer Services, and Cellar Services, received a rating of Excellent under the Building Research Establishment Environmental Assessment Method (BREEAM) system on completion, and attests to the fact that an environmentally friendly building is able to meet the financial and operational requirements of such a company. The structure has also won the David Alsop Sustainability Award 2007 for achievement in structural design, with outstanding commitment to environmental sustainability, and was nominated for the 2007 RICS East of England Award for Sustainability.

The investment by Adnams, both in the fundamental building technology and the actual construction of such an innovative building, has also been pivotal beyond its own environmental achievements. Adnams gets a very high number of visits from people inspired by the building who come to discover firsthand what has been done, what efficiencies have been achieved, and what could have been done differently. Over the years, these have included representatives of businesses, the UK government (including Department for Environment, Food and Rural Affairs), and the US government. Adnams representatives have also been very happy to accept invitations to other organizations to talk about the sustainability initiatives. As a result, while a new enterprise center at the University of East Anglia and a new spar at Coworth Park Spar in Berkshire will be constructed from the *Tradical* Hemcrete material, a wealth of the other techniques have been employed in constructions by retailers such as Marks & Spencer, Tesco, and

Sainsbury's. As Andy Wood explains, "We are not claiming credit for what others have done; we are just saying that we had it here and people have come and looked at it, and people have taken that on a stage further, and that's a good thing because that means buildings are even more sustainable. I don't see this as a competitive issue at all."

Having said this, there are still many areas for improvement within Adnams. For example, the company has not engaged in any active waste management at the site. Additionally, the distribution center is yet to deal with latent operational impediments principally caused by the use of different standards of palletization across different ranges of beers and wines. This represents an inhibitor to the performance of the distribution center and still generates storage inconveniences, which have their own environmental impacts.

East Green Beer: A Celebration of Environmental Values

After the brewstream and distribution functions of the Adnams business had been addressed, Andy Woods spoke at a seminar on issues of sustainability in the brewing industry. Also speaking at the event was the director of Beer, Wines, and Spirits at Tesco, who raised interest in working together to develop a *carbon-neutral* beer as a natural extension of Adnams' existing activities. The team at Adnams began to work with Tesco, as well as the University of East Anglia's CRed carbon reduction scheme and the Carbon Trust, to leverage existing gains and produce Europe's first certified carbon-neutral beer, with 25% less carbon compared to Adnams other products.

The Adnams team thought the obvious place to look for additional carbon savings was the glass packaging of the product, as by far the most carbon-emitting part of the process is the blowing of the glass. In response, Adnams approached their bottle manufacturer, OI, and through working together, managed to achieve a 33% reduction in weight, according to analysis by the University of East Anglia.

Another part of the supply chain that Adnams reviewed in order to produce their carbon-neutral beer was the supply of grain and malt. To reduce the carbon output from transport, Adnams sources grain from the twelve farms that are closest to their maltsters (who make barley grain into malt by wetting corns to make them grow, stopping the process with heat at the right point to soften the grain and release the flavor). The company also worked with the Suffolk Hop Association to naturally breed a hop that was aphid resistant and, therefore, did not require the oil-based application of herbicides or pesticides.

After these processes, there is still around 142 g of carbon per bottle produced—about the equivalent of traveling 1 km in an average sedan car. The remaining carbon output is offset by Adnams through a gold standard offset scheme run by ClimateCare. Although this is priced at 0.004 p per bottle, Adnams pays 1 p for each bottle sold. In recognition of this overall achievement, the UK's Carbon Trust, which had just bestowed their Carbon Innovator of the Year award on the company, allowed Adnams to include their logo on the bottle.

While this string of innovations has resulted in Europe's first certified carbon-neutral bottled beer, more importantly, the lighter packaging enables Adnams to move more bottles on a lorry, making distribution more efficient. Andy estimates that this initiative saves something in the region of 700 tons of carbon a year, or more imaginatively, the carbon equivalent of not only all staff journeys to work, but also all the journeys undertaken by Adnams sales personnel.

Adnams does not bottle its own beers, partly because bottling equipment is very expensive, and partly because the current bottler based in the Midlands has already made substantial investments in modernizing its own process. While this might lead to the suggestion that extra environmental costs are being incurred, Adnams is satisfied that this is the most eco-efficient arrangement currently available. This is because under the current arrangement, beer is moved in bulk tankers to the bottling facility, which is conveniently located very close to a number of regional distribution centers for large supermarket chains, to which bottles are transported as part of wider consignments.

Despite East Green being a tremendously significant environmental achievement, this is not the end of the story as far as the company is concerned. Indeed, asked if the current need to use geographically proximate farmers to produce East Green beer limits what can ultimately be achieved, Andy responds, "Absolutely not. We are not at an end point. This is part of a journey that we are on. We are absolutely not at an end point." One such area of possible future continuous improvement is in the production of beers with a more international flavor. For example, during the Rugby World Cup, the brewery produced a themed beer flavored with hops from New Zealand. However, with an eye to sustainability, Belinda, the quality control manager, says that the company is again working with the National Hop Association of England to produce the same flavors from hops grown in the UK, and it is hoped that Adnams will be able to source the ingredients at a much reduced financial and carbon expenditure in the future.

Perhaps the most significant impact of Adnams, however, has been that the environmental gains pioneered in the development of the East Green beer have not stopped with the company. Indeed, the lightweight bottles developed under the brewer's leadership have quickly become an industry standard. As Andy describes, "Not only did we make a saving of 700 tons of carbon per annum, but actually there was a saving right across the industry. We had the thought leadership around this and others followed suit." And again Adnams is not content to let the innovations stop there.

Bio-Digester: An Industrial Ecology

Given that the sustainability activities are driven by a genuine belief in the importance and good business sense of green operation, it is not surprising that Adnams has continued to innovate even after the recognition of its carbon-neutral product line. First and foremost, the company is currently working with the University of East Anglia to produce carbon and water footprints for all its beers as a means to locate further opportunities to drive down the company's overall carbon output.

Aside from this, it is noted that the next step on the journey was to look at the waste. As a result, Adnams has worked with the Cambridge-based Bio Group, and launched a joint venture company with them, Adnams Bio-Energy. At the heart of this initiative, Adnams installed a 12,500-ton anaerobic digestion plant in 2010. The facility uses micro-organisms to break down biodegradable material, or act as "a mechanical cow's stomach" that allows the company to extend their use of industrial ecology, where waste from one part of the production process becomes energy for another part.

This initiative, once again a first for a UK-based brewery, takes *ullage* (or waste product) from the bottom of brewing tanks (which accounts for about 4,000 tones of capacity), food waste from the Adnams pubs and hotels, as well as other food waste from within a 25-mile radius, to produce bio-methane. Forever striving for innovation, the engineering of the digester offers improvements over existing types (using air and gas to agitate the material rather than a mechanical process) and is the UK's first bio-digester to supply food waste produced gas into the national grid. Indeed, working at full capacity, the digester can supply enough gas to generate up to 4.8 million kilowatt-hours, or enough to heat 235 family homes for a year.

However, with the purchase of a bio-methane filling station, Adnams plans to acquire their first two dual-fuel trucks from Volvo in 2012. Indeed, the long-term plan is that over the next 4 to 5 years, the company will replace its entire commercial fleet, for a further carbon saving of 30% off its overall remaining footprint. The final part of this strategy will see the supply of by-products from the digestion process, *leechate* and *digestate* (described affectionately as "baby-bio on steroids"), free of charge to the farms that grow Adnams inputs. In this way, Adnams management says, the production loop will be closed and another contribution to industrial ecology will be made.

Incremental Sustainability Initiatives

While many big environmental improvements have come through making fundamental changes to business infrastructure and large-scale investment, the company is also proud of a plethora of smaller incremental environmental initiatives. For example:

- When Adnams added spirit distillation to their portfolio of business activities, they ensured that the design of the equipment captured heat from the condensing process to supply the water used for washing and sterilizing the ale casks in the casking room below.
- The company has rented land to British Gas for a nominal fee to set up solar panels. The site will supply Adnams with the electricity it needs and then sell the rest back to the grid.
- The cafe within the Cellar & Kitchen store has a seating area recycled from copper brewing vessels and old beach groynes.
- Adnams' hotels and pubs have replaced gas heaters with the distribution of blankets to keep customers warm in outside spaces as a means to reduce energy consumption.

It should be remembered, however, that despite the emphasis on environmental improvements, the motivation of these initiatives comes from a community-embedded feeling of social responsibility— in terms of both future generations and contemporary stakeholders in Adnams' operation.

As such, Adnams seeks to be a wider agent of change by promoting their own company values, working with like-minded companies, and seeking to facilitate change in others where possible. This is not enforced in a standardized list of criteria—although the one absolute prohibition is on the award of supply contracts to companies that use child labor—but through a consideration as to whether potential contractors have compatible values. Adnams will give new contractors a chance, the operations director explains, but "we want to work with like-minded people. We also like to help people do the right thing."

Practicing what they preach, Adnams ensures that pay levels for all its employees are benchmarked by an independent company that standardized the criteria, locating the lowest, median, and highest pay levels available in the region. Adnams then pays the median and offers all staff a 4% addition to their pay in company shares.

This sense of distributional justice is widespread, and in 1990 the company founded the Adnams Charity to celebrate its centenary as a public company, toward which at least 1% of the profits of Adnams PLC. are put each year along with other private donations. In 2009–2010, the trust awarded grants totaling £76,463 to 61 good causes within 25 miles of St. Edmund's Church in Southwold. These included local schools and colleges, the St. John's Housing Trust, the Fisher Youth Theatre Group, and Westleton Hall, among many others.

Finally, Adnams management feels that the ultimate sustainability and continuous improvement initiative revolves around the investment in their workforce. For example, the company's development ladder sends staff away to undertake additional training, right up to the level of an MBA if they so wish. Also, the management team is a mixture of those brought in from the outside and homegrown talent, the story of Karen Hester being the most pronounced example (see Box 2). For this reason, there is a high level of mutuality throughout the company, which is significantly assisted by the fact that in many cases, managers rarely ask the workforce to engage in things that they have not themselves done at one time or another.

Business Gains from Sustainability: The Lean Side of Green

While the management team at Adnams offers an exemplary example of green innovation and radical thought leadership, representatives are also clear that they are still a PLC enterprise. The operations director comments, "We are a business after all, we've got shareholders and we have to justify our expenditure. It's always got to make business sense, because we are in business, and wouldn't be if we never took that into account. So first and foremost, everything has to make business sense."

In these terms, it is pertinent that the new brewing equipment installed in 2007 came at a premium over the cost of a more conventional replacement. However, as there were some delays with the replacement project, trouble in the international economy saw the cost of steel increase. As a result, the premium for the eco-efficient brewstream came down 30–40% to something more in the region of 15%. Likewise, the additional cost required to build the eco-distribution center was 15% to 20% in comparison to a standard steel-framed warehouse of a same size.[2] In the words of Andy Wood, Adnams has "some big blocks of shareholders and it was important to convince them that this was the right move to make."

However, when the subsequent running cost of the eco-efficient infrastructure is compared to a conventional design, the business case for such investments has been overwhelming. The new distribution center uses 58% less gas and 67% less electricity per square meter than the old structure. Therefore, based on 2004 fuel costs (when Adnams board made the decision to build the eco-distribution center), savings have been £50,000 per annum. Where the calculation is based on 2006 fuel prices, savings are in the region of £49,000 per annum. Karen Hester estimates that if Adnams would have to cool their current volumes of ale, it would cost in the region of £100,000 a year in utility costs. In fact, the energy bills are actually the same in 2011 as before the company left the old center, despite a significant growth in their output.

As a result of these savings, it is calculated that the cost to Adnams to serve its customers is the same or less in 2011 than it was before the innovations were introduced, and furthermore the company has been able to hold its beer prices for three years, which according to management, "is just more proof that the investments made in sustainability were the right thing to do." While this has naturally been a socially supportive thing to do for tenants struggling to deal with a wider economic downturn, spreading this strain has ensured that businesses remain open and Adnams maintains its outlets and market share. Indeed, Adnams has doubled beer sales volume since 1997, despite a market that has been in continual decline, falling by 8.1% in 2010 alone.

Aside from these savings, there is a clear feeling at Adnams that the green element of their brand adds concrete value to the company. As a small company, Adnams does not have a huge marketing budget. However, according to a company representative, "The green element to our voice has created interest in the company and has created cut through for the brand. We have everybody beating a path to our door, from Korean TV to CNN, who spent a whole day filming here when we opened the new brewery." As Andy Wood explains, competition for supermarket shelf space is usually particularly intense, especially as retailers use beer to attract consumers into their stores and look upon it as a commodity product. However, the distinctive nature of the East Green line places Adnams in a different position given their partnership with Tesco, as the brewer was seen as an added-value supplier and an innovator, by both them and other supermarket chains.

When asked if he thinks companies that do the right thing are more financially successful, Andy Wood replied, "I think the jury is out on that one. There is research emerging that suggests, over the long term, companies that do the right things will be more economically successful. We have always tried to deliver sustainable business results for our shareholders. That is particularly important for our shareholders."

This does not mean that everything has been smooth sailing for Adnams. While revenue increased from £46 million in 2006 to over £51 million in 2009, this declined slightly in 2010 to just under £51 million, mainly as a result of some of their bigger customers being impacted by the financial downturn. This also contributed to a drop in profitability in 2008. Indeed, the recent story of Adnams has been played out against a wider backdrop of serious economic contraction across the whole of the UK pub industry, at the nadir of which, some 50 pubs closed every week.

However, supported by the concrete savings from the lean and green business approach, Adnams' Annual Report of 2010 shows that company turnover has risen to £54.5 million, a healthy 7.2% year-on-year increase from 2010, while the operating profit also has risen by 2.8%.[3] In order to reinforce the sustainability of this recovery, the company has been making a proactive effort to reorient its sales away from a small number of large customers and into new and more diversified distribution channels. The past years have been a learning curve and, says Andy, "We will be a more resilient business for the experience." Moreover, share price has been held constant, even at times of lower profitability, and as noted by company representatives, "People tend not to sell their Adnams shares, but instead tend to pass them down from one generation to the next." This broad appeal of the brewery is also manifest in the fact that in 2010 Adnams won the Brewery of the Year competition run by the UK's bestselling *Good Pub Guide* as voted for by everyday customers. Indeed, if Adnams can continue to expand its basic customer loyalty, this is another essential component in the development of long-term business sustainability.

Conclusion: Innovative Solutions and Human Drivers

The case of Adnams is without doubt an outstanding example of green business activity in that in the last decade, the company has considerably reduced its impact on the wider physical environment and increased its positive social and economic returns. The company has become more and more profitable (leaner) during this period, wasting less, producing more with less energy and resources, becoming more resource efficient, and most importantly, becoming more effective by enhancing its brand and going beyond customer and consumer expectations.

One of the primary elements in this story has been the innovative use of genuinely cutting-edge green solutions such as the installation of one of the most energy and raw material efficient brewstreams in Europe, construction of an eco-efficient distribution center, and setting up an anaerobic digester that takes Adnams waste from brewing, retailing, and pubs and turns it into biogas feeding the national grid. While these initiatives have been successful in themselves, perhaps a more important element has been the way that leadership is thinking and doing things differently providing concrete examples of what can be achieved in the food and retail industry and even beyond.

The willingness of Adnams to be early adopters of solutions that may even seem expensive at the beginning, has left a significant positive footprint on the intellectual and physical landscape of the sector. It also proves that going green pays off economically. Closing the energy and material loops in Adnams, for example, by turning waste material into biogas or reusing heat in their modern brewing process, has certainly saved a lot of waste turning Adnams into a very lean company.

Arguably, the most significant part of the Adnams story is the human element that drives the application of these ideas. As can be seen from the combination of the main narrative and personal stories in the text boxes, individuals play key roles in developing the Adnams way, although we have given but a few examples of how committed staff is across the whole company and across different functions.

The key features of lean and green leadership at Adnams are as follows:

- Value-based decision making with a deep conviction that going green pays off
- Engagement
- Tenacity
- Proximity to workplace

In terms of company *values*, similar to Toyota, the company has had a longstanding sense of wider responsibility that has developed in the DNA of the organization. Adnams, like Toyota, does not see itself as a short-term money-making machine, but rather a responsible member of the society serving the society's desire for responsible drinking and pleasure. Here, strong and centralized family ownership has been, and continues to be, vital. However, one of the major influences in the development and formalization of this background was the entrance of the current CEO, Andy Wood, whose former experience (and sometimes frustrations) in developing a values-led approach to business gave the organization clear and inspirational leadership. This approach to business offers an archetypal example of lean and green business leadership through a combination of economically motivated business operations and an inherent commitment to environmental sustainability. The key here is that these two often juxtaposed perspectives are reconciled through a non-zero-sum perspective, which is the recognition and the deep conviction that what is good for the environment in the short term is also good for the development and sustainability of business in the longer term.

In terms of staff *engagement*, while strong leadership manifests itself in the top-down implementation of some major (sometimes technologically grounded) changes, the management has also recognized the benefits of genuinely embedding values across the whole organization. Today, Adnams is committed to environmental objectives from the bottom up as much as it is from the top down. Introducing recycled blankets in Adnams pubs, led by frontline employees, instead of the usual gas heater solution is a good example of bottom-up engagement across Adnams. Creating this level of commitment has taken time and has required persistence. The creative use of standard management techniques, such as eco-champions and performance-related pay, and the development of more novel systems, such as confidential staff survey and regular "snack and chat" sessions with staff, has had a significant influence on the everyday behavior of Adnams' workforce. This team-orientated approach is also evocative of the lean system as it initially developed in the manufacturing sector in Japan.

Tenacity is another key facet of leadership in Adnams.

You may remember this simple yet startling high school physics experiment. The student runs a metal paperclip across a magnet repeatedly, always stroking in the same direction. After maybe 100 or 200 strokes, the paperclip has obtained magnetic properties and can attract other metals or stick onto the fridge door. However, should the experimenter stroke the paperclip in different directions, or should they make a back and forth motion with their strokes, the paperclip will not become magnetic. What the experimenter is doing is aligning the atoms inside the paperclip in the same direction by repeatedly stroking it with another magnetic field. As soon as the experimenter changes direction, the magnetic properties of the paperclip, which are not permanent yet, will be lost. Even worse, if the student drops the paperclip amid the experiment, it will lose its magnetic field immediately. The experiment is analogous to leadership that equally requires great persistence and constancy of direction. In order to make sure the lean and green vision and values sink into every individual within the organization, the leadership team at Adnams has been pursuing sustainability with a great deal of tenacity, making sure that short-term considerations do not influence the company's direction and that even the most skeptical people are slowly brought round.

Regarding connection *to workplace*, we showed how the company stays connected and familiar with the frontline operations. Several members of the leadership team at Adnams have worked their way up the company ladder from the shop floor, including the existing operations director, Karen Hester, who started in the company as a part-time cleaner. Adnams' determination to stay close to the workplace and its desire to understand staff needs are reminiscent of Toyota's Genchi Genbutsu (go to workplace and see) principle.

Having said all this, and despite the already tremendous achievements of Adnams, there is still potential for more improvement. For example, the consumer side of the supply chain still has to be tackled, where more attention can be paid to wastage and CO_2 emissions generated in the consumption process. Likewise, while Adnams has undertaken initiatives to close the production loop with suppliers (by partnering on the development of new hop varieties and investing in bio-fertilizer production), there is still great potential for continuous improvement. Moreover, Adnams only started on its systematic lean journey less than a year ago. Although, continuous improvement for green and resource efficiency already existed within Adnams, they did not have a formal lean office, which they have now identified as an opportunity since around a year ago. The aim is to benefit from the potentials that systematic lean methodologies can offer. Establishment of a formal lean office has been fairly straightforward due to high levels of bottom-up engagement. lean improvements have started in manual operations such as the warehouse and have been evolving since. It is interesting that Adnams is one of few companies that were introduced into the domain of lean through green continuous improvement.

It is here that Adnams must continue to prove itself, if the lashings of the North Sea are to be stabilized, and mitigated, for the good of present and for future human welfare.

Endnote

1. Statements belonging to Adnams employees were collected by authors through various interviews.
2. Suffolk; UK n.d.
3. Adnams (2011). *Adnams Annual Report and Accounts*, Suffolk: UK 2011. http://adnams-misc.53.umazonaws.com/randa-2011.pdf.

Chapter 6

Lean and Green
Strategy Deployment

In this chapter we focus on the *strategy deployment* element of the Lean and Green Business System model (Figure 6.1). We present a case study of Tesco Plc. for three simple reasons: Tesco is a landmark example for successful strategy deployment, Tesco is a pioneer of lean operation across the end-to-end processes, and they have been leading the way in greening the industry. But what underpins Tesco's success in both lean and green is their remarkable ability in dynamic strategy formation and strategy deployment. Over the past two decades, Tesco has demonstrated an incredible capability to create a highly aligned and engaged organization. Tesco refines their strategy based on the changing needs of the markets to stay in tune with their core purpose. Having clear purpose, vision, values, and strategy is important to Tesco's landslide victories, but what really underpins their success is their ability to align and engage the entire organization with their values and strategy.

Lean and Green Strategy Deployment:
Case Study of Tesco Plc[1]

A Brief History

Since it was founded in 1919 by Sir Jack Cohen as a group of market stalls, Tesco was destined to dominate in the global retail landscape. In two decades, Tesco numbered over 100 stores across the country, and while

Figure 6.1   Lean and Green Business System model.

Sainsbury's made a niche through quality, Tesco's focus was on keeping prices low. The *pile it high and sell it cheap* approach was particularly attractive for British consumers in the postwar 1950s. During the 1950s and 1960s, Tesco grew steadily, and also through acquisitions, until it owned more than 800 stores. The company purchased 70 Williamson's stores (1957), 200 Harrow Stores outlets (1959), 212 Irwins stores (1960), 97 Charles Phillips stores (1964), and the Victor Value chain (1968; sold to Bejam in 1986). As Tesco was focused on price, the goods available at the stores were perceived to be of run of the mill quality, but when the country started to overcome the recession, baby boomers changed from price to quality conscious, and as the national average income rose, customers began to look for more expensive and luxury merchandise. This led to the rise of a more premium British retailer J. Sainsbury's. In the early 1990s, the threat of losing position in the UK market led Tesco to establish strategic values that would later become a landmark in global retail history. From humble beginnings to world domination, as Tesco's success story continues, what remains at the heart of their strategic values is a relentless focus on the needs of the customers.

Today, Tesco is the third largest retailer on Earth with just under $100 billion in revenue, more than 5,000 retail stores across 14 markets and

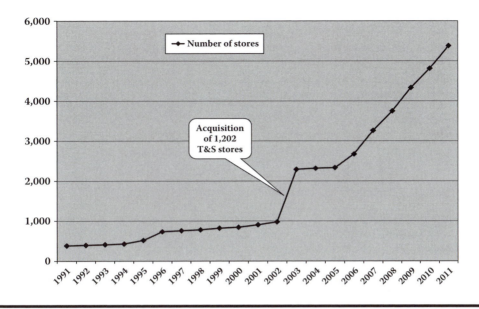

Figure 6.2   Growth of Tesco from 1991 to 2011. (Data from Tesco website and other publically available sources.)

nearly half a million employees. Tesco's growth and success is illustrated in Figures 6.2 through 6.4, which track their volume and profitability over two decades from an average UK retailer to a major global player owning 2,715 stores in the UK and an additional 2,750 worldwide. Interestingly, for the last 20 years, Tesco has kept its profit margins constant, fluctuating just

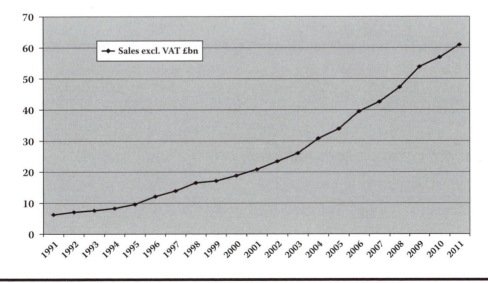

Figure 6.3   Tesco sales (excluding VAT) in £ billion from 1991 to 2011. (Data from Tesco corporate reports.)

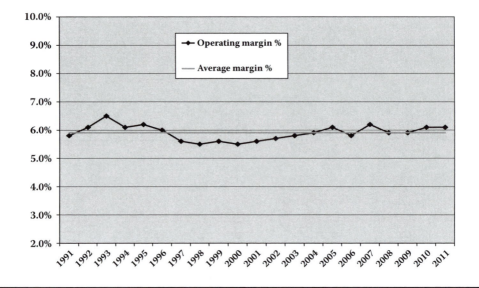

Figure 6.4 Tesco operating profit margin from 1991 to 2011. (Data from Tesco corporate reports.)

around 6%.[2] This healthy growth and profitability is no coincidence, but the direct result of their remarkable *constancy of purpose*. A business that has grown, by and large organically, by learning to become leaner every day and reinvesting some of the short-term gains in future competitiveness while keeping prices low and profits constant at a healthy level. Tesco is a lean company. In fact, Tesco has been actively adopting the principles of lean thinking since the early 1990s. In a public seminar in 2010, Sir Terry Leahy, CEO of Tesco at the time, revealed that lean thinking has been one of the key management principles guiding him and Tesco during their meteoric rise.[3] Achieving what Tesco has achieved in a mature sector with seemingly few opportunities for innovation is truly exceptional. In this case study, we explain some of the key attributes that make Tesco so different.

Tesco: A Customer-Focused Company

It is notable that what differentiates Tesco is certainly not in adopting a unique business model. They follow the same business model of *buy, pile, and sell in store* as any other large or small retailer. One might, then, argue that achieving such colossal success emanates from an inimitable strategy. However in fact, Tesco's strategy is nothing exceptional. A BBC documentary described them as "being everywhere and selling everything to everyone."[4] Clearly, this strategy per se is not and cannot be a differentiator for Tesco.

In Tesco's case it is the way in which they execute their strategy that sets them apart from the competition. Tesco has been setting audacious goals while also establishing strong strategic vision, values, and culture. Having a very clearly defined and well-established *core purpose* creates clarity for the company and allows Tesco people to be highly focused and aligned in everything they do.

But what is Tesco's core purpose? One of us often delivers workshops and puts Figures 6.2 to 6.4 on a slide for the audience, sometimes an auditorium full of managers, asking them what they think Tesco's core purpose is? The answer often comes back in phrases such as growth and profitability; even world dominance or Tesco's own advertising motto *"Every little helps."* The truth is that none of these terms appear in Tesco's core purpose.

Their core purpose is *to create value for customers and to earn their lifetime loyalty.*[5] This makes Tesco axiomatically a lean and green organization at a corporate level. Factors such as growth and profit are treated as internal or *lag* measures of success; they are important, but subordinate to achieving the core purpose. In the past two decades, Tesco has taken great care to create a highly focused and aligned organization articulating the core purpose and Tesco values as well as a clear vision and strategy. They go to great lengths to communicate the core purpose and to engage at all levels. One of us has been as far as Thailand observing the core purpose and Tesco values being visibly communicated with all employees. What underpins the company vision and strategy is still their core purpose and a highly focused organization pursuing it.

While Tesco's values and core purpose have remained unchanged since they were established in the early 1990s, there have been some changes to the company strategy to reflect the ever-changing needs of the customers. Most notably in 2007, the company added a fifth objective to the then four-part strategy to underpin Tesco's commitment to becoming *truly sustainable.* More recently, in March 2011, when Philip Clarke replaced Sir Terry Leahy as the new CEO, he added two new objectives to create the new seven-part strategy illustrated in the Figure 6.5.[6]

Clearly, the point here is not Tesco's ability to come up with nice words to describe their values or strategy, but rather their incredible ability to create a highly aligned and engaged organization as well as their ability occasionally to refine the strategy based on the changing needs of the markets to stay in tune with their core purpose. Again, Tesco's uniqueness is not in strategy formation, but rather strategy deployment as we will explain in the following. No matter how clever your strategy is, it cannot be deployed

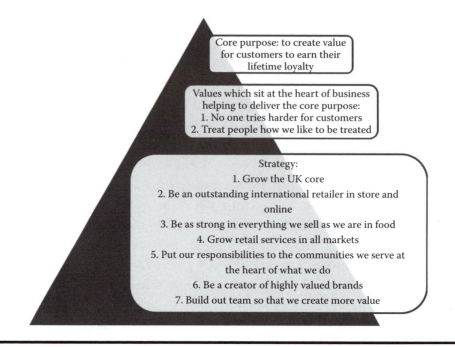

Figure 6.5   Tesco's core purpose, values, and strategy as of April 2012.

successfully without getting everyone in your organization involved and making them aware of what the business direction is and what their contribution will be in its achievement.

There are various methods by means of which Tesco ensures all its activities are aligned to its core purpose. For instance, the strategy and the core purpose are streamed down to the operational level through the triptych of *better, simpler, and cheaper*—being better for customers, simpler for staff, and cheaper for the organization.[7] Tesco applies better, simpler, cheaper to make sure every single project and activity is aligned to the core purpose of the company. For example, in considering operational improvements or business change, impacts are assessed against better, simpler, cheaper principles, where all three of them must be satisfied—not just one or two, but all three.

But Tesco has an even more powerful tool for strategy deployment embedded throughout the organization. The Tesco Steering Wheel, introduced in 1996, enables strategic management to be conducted in real time by setting up local and individual targets and building appropriate measures to quantify performance against those targets. Tesco's core purpose, values, vision, and strategy are the foundations for the Steering Wheel, which in turn drives the day-to-day business and prioritizes the key issues.

The steering wheel is a simple yet effective tool that drives high levels of *alignment and engagement*; it is intended not so much to monitor staff performance and target fulfillment, but to drive behavior and help staff to do the right things. Since the addition of the sustainability segment to the Steering Wheel in 2007, there are five parts monitored in the wheel: customer, people, operations, finance, and community, which reflect the sustainability measures.[8]

Each segment in turn is broken down into smaller sections with their respective sets of key performance indicators (KPIs) based on challenging yet realistic targets. Quarterly performance reports are sent to the board, and a summary report for the top 2000 managers worldwide is meant to cascade to staff. The senior management pay is linked to the KPIs with bonuses being based on the level of achievement on the Steering Wheel.[9]

Figure 6.6 illustrates the Steering Wheel. The wheel is typically updated weekly and performance is usually reported with different colors, where green denotes acceptable performance, amber is for borderline performance, and red for unacceptable performance.[10] Not only are Tesco managers measured according to their performance against the Steering Wheel, but different functions and stores have their own Steering Wheels reflecting the targets that are most relevant to them. This means that a store manager

Figure 6.6   The Tesco Steering Wheel after 2007. (From J. Tippett, *Informing Choice, Leading Change: Towards a Sustainable Future*, Manchester, UK: University of Manchester Sustainable Consumption Institute, 2009.)

in Turkey or Taiwan will have similar responsibilities to one in London. Moreover, each year, KPIs are set in each of the fourteen countries where Tesco operates in order to measure projects targets as well as operational performance.[11]

In January 2007, Tesco evolved its strategy to embrace *true sustainability*. This was marked by the CEO's speech to Forum for the Future making clear commitments on behalf of the company:[12]

- Tesco will be an active leader in helping to create a low-carbon economy (as opposed to being a passive follower).
- Tesco will measure and reduce greenhouse gas emissions. Thus it would set stretching targets, committing to these in public.
- Tesco will help to stimulate the development of low-carbon technology—for example, in the construction and operation of stores, distribution centers as well as in its supply chain. This would include working with suppliers and others to deliver significant CO_2 reductions.
- Tesco will use its unique relationship with customers to help deliver a revolution in green consumption with the fight against climate change at its heart.

In the same speech Leahy said, "We are going to have to re-think the way we live and work … the green movement must become a mass movement in green consumption." Ever since that time, Tesco has been embedding these commitments into their business strategy via their Steering Wheel mechanism. The latest part of the Steering Wheel, Community, emphasizes the corporate social responsibility profile of the company as illustrated in Table 6.1.

Within the *Community* segment, Tesco makes five key Community Promises:[13]

1. Buying and selling our products responsibly
2. Caring for the environment
3. Actively supporting local communities
4. Giving customers healthy choices
5. Creating good jobs and careers

Like other parts of the Steering Wheel, Tesco set clear targets and KPIs for monitoring each of their Community promises. Table 6.1 illustrates some of their latest KPIs as of the beginning of 2012, as well as their performance over three years.

Table 6.1   Sample of Tesco KPIs for Monitoring Community Promises

Community Promise	Key Performance Indicators	Performance (%)		
		2008–9	2009–10	2010–11
Buying and selling our products responsibly	Supplier Viewpoint: average score (% of scores that are positive)	68.0	80.0	82.0
	Supplier Viewpoint: response rate of suppliers	37.0	51.0	54.0
Caring for the environment	Reduce CO_2e emissions from our 2006–7 baseline portfolio of stores and distribution centers by 50% by 2020. Annual target reported as percentage reduction against previous year.	7.0	7.8	7.7
	Reduce CO_2e emissions from new stores and distribution centers built after 2006 by 50% by 2020, compared to new stores and distribution centers built in 2006.	20.5	28.8	28.8
	Reduce the amount of CO_2 used in our distribution network to deliver a case of goods by 50% by 2012, compared to 2006. Annual target reported as percentage reduction against previous year.	9.2 (UK)	6.4	7.4
Providing customers with healthy choices	Staff and customers active with Tesco.	4.7 million people	6.2 million people	Over 7 million people

Source: Collated from *Tesco Corporate Responsibility Report*, 2011.

Thus, the Steering Wheel monitors how well the company keeps its Community promises. It quantifies the three pillars of sustainability—*economy, society, and environment*—which are considered and embedded in Tesco strategy under the main guidelines to *be responsible, fair, and honest,* and *be a good neighbor.*

Tesco managers are equally measured according to their performance against the Community segment. For example, each store will have to report on carbon and waste reduction targets as well as the other Community initiatives. Similarly, managers in every market will be contributing to Tesco's charitable giving targets and helping to get people active. A more extensive list of various performance indicators used to measure sustainability performance across Tesco stores is given in Table 6.2.

While an instinctive focus on customer satisfaction has been at the heart of Tesco's business for over two decades, their understanding of the needs of shoppers has been evolving constantly. Tesco has gone through various stages of maturity in refining their internal emphasis to *create value for customers to earn their lifetime loyalty.* They started acting on the *internal efficiency* of various parts of their process in the early 1990s, but soon learned that true competitiveness is in *end-to-end efficiency* along the whole supply chain from the supplier's factory gate to the retail shelf. Tesco triggered a lean program to extend efficiency to the whole chain in 1996[14] and introduced the company's balance scorecard, the Steering Wheel, the same year, which also reflects the whole chain efficiency thinking. By the year 2001, they had reported the equivalent of £150 million annualized savings as a result of applying lean solutions to their end-to-end supply chain, such as continuous replenishment, among other things.[15] These benefits were delivered under the *Step Change* program, which covered the triptych of "better, simpler, cheaper," and have continued to date. Step Change was Tesco's version of continuous improvement or lean. By 2010, Tesco's ability to create benefits and savings under the Step Change program had risen to £550 million per annum.[16]

In fact, Tesco's business is about running an efficient supply chain, as they don't make anything themselves. We explain some of the key innovative lean solutions, such as the one-touch replenishment concept, flow-through supply chain management, and factory gate pricing that set Tesco apart from the competition during 1990s and 2000s in the following sections.

But at the same time that Tesco began working on its end-to-end efficiency, there was another important development occurring within the marketing department. Throughout 1994, Tesco carried out a series of trials

Table 6.2   Examples of Sustainability Measure across Tesco Stores

Sustainability Pillars	Selected Examples of Performance Categories Being Monitored	Selected Examples of Key Performance Indicators Deployed
Economy	Regeneration	Number of jobs created
	Supply Chain labor standards (SCLS)	Number of ethical assessments carried to high-risk own-brand suppliers
	International Corporate Responsibility (CR)	Number of international businesses to have a CR strategy with performance measured in their Steering Wheel
Society	Charitable giving	Percentage of pretax profits donated to charities
	Computers for schools	Cumulative value of computer equipment supplied to schools in million pounds sterling (£M).
	Employee retention and training	Percentage of retention of experienced staff—Percentage of retail saff to be trained to Bronze level
	Inclusivity and diversity	Statistical difference by age, sex, or ethnicity in answer to the staff viewpoint Survey question: "I look forward to coming to work"
Environment	Energy efficiency	Reduction of energy consumption measured in kwh/ft^2 and as a percentage of total consumption
	Water efficiency	Reduction of water consumption measured in m^3/m^2, as a percentage of total consumption and as reduction in m^3 of water per annum
	Transport efficiency	Percentage increase in the volume of products delivered per liter of fuel consumed
	Landfill avoidance	Percentage of waste recycled

Source: Adapted from S. Pradhan, *Retailing Management: Text & Cases.* (New Delhi: Tata McGraw Hill, 2009).

with a new loyalty card scheme. When shown the results of loyalty card's trials, the then-chairman, Lord MacLaurin, is reported to have said: "What scares me about this is that you know more about my customers after three months than I know after 30 years."[17] Shortly after, in 1995, Tesco launched its first-ever electronic loyalty card program, initially called the *Thank You Card*, but later redubbed the *Tesco Clubcard* program.

This was a move to better understand their consumers and to reward loyal shoppers by giving back 1% of the sales value to the shoppers in the form of money-off vouchers. It later became an integral part of a move by Tesco toward an even more customer-centered strategy; the era of end-to-end effectiveness had begun at Tesco. In spite of high setup costs, and in spite of the fact that the 1% payback amounted to nearly 20% of Tesco's profit at the time,[18] they decided to do it. This was a risk that Tesco was willing to take as the scheme allowed the company to develop intimate one-to-one relationships with individual shoppers, taking their consumer centricity to a different level.

Soon the Clubcard was an important tool for gathering personal information about what consumers buy, how much they spend, how often they shop, and when they shop, with the information revealing a great deal about shoppers' lifestyles and individual habits. Within a few months, five million people had signed up and as of 2011, it had over 16 million members (or approximately 25% of the total population of the UK). This rich source of information is the best way for the company to get to know its customers. Moreover, the Clubcard program generated in excess of £53 million profit in 2011 by selling the (anonymous) consumer data to third-party companies eager to understand what customers really want.[19]

Moreover, it helps Tesco avoid huge unnecessary costs by only stocking products that will sell. Combining the end-to-end supply chain efficiency solutions with effectiveness solutions and consumer centricity has allowed Tesco to become a truly lean company over in the last twenty years. What underpins lean solutions and the Step Change program at Tesco is their corporate culture driven by strong company values and a clearly defined core purpose. At the same time, a well-deployed strategy, enabled by the clarity provided by the Steering Wheel, gives Tesco the necessary focus to deliver the core purpose.

Tesco sees its supply chain as a factory delivering convenience to shoppers. In order to earn customers' lifetime loyalty, they focus on exactly what consumers want, and for this purpose, they have the ideal weapon. The Clubcard program has given Tesco the unique ability to tailor the range of products listed in each individual store to the shopping profiles of that store's customers. Moreover, Tesco was a pioneer in adopting multitier product ranges to ensure they would cater to all different incomes and tastes. For private-label products, across a given category, they offer three product tiers—Value, Standard, and Finest[20]—as well as subbrands such as Healthy Living, Tesco Kids, and Organics.

Also they deploy the Clubcard data to extend beyond traditional retailing into areas such as personal financial services (currently 2.5 million customers). Since the year 2000, Tesco has launched into the online retail world, quickly becoming one of the largest global food retailers online. Online shopping or Tesco.com provided them with yet another insight, this time showing them exactly what their in-store availability levels are because for the first time they discovered what consumers wanted to purchase but couldn't.[21] Tesco decided to fulfill online shopping orders using their store staff at times when they were not very busy and to eliminate the need for separate expensive depots and fulfillment centers for home shopping. The dot-com availability measure is very important for Tesco as a rich and reliable way to understand their on-shelf availability, a key indicator of shopper satisfaction and something about which Tesco has been obsessed.

The way Tesco thinks about on-shelf availability is not just at an individual product item level, but in terms of *basket fulfillment*. For example, while 98.5% availability for most product items is a very good industry average, it translates to a much lower basket fulfillment of only a 55% chance to find all items for a basket of 40 products.[22] The main objective in a consumer-driven supply chain is to deliver convenience, to aim to have all products 100% available so as to earn customer loyalty. Tesco combines its consumer insights obtained through the Clubcard program and Tesco.com with innovative lean solutions to build a uniquely resilient yet flexible business.

In 2005, each square foot in Tesco's yielded £22 a week, compared to £17 at Sainsbury's and only £10 at Marks & Spencer.[23] This high sale density could only be achieved by keeping the shelves well stocked and making sure that the whole supply chain was working rapidly and efficiently. Supply chain speed and accuracy are key. Products have to be delivered from their point of origin to the points of sale with a minimum of effort, minimum error, little waste, and no delays. One of the key lean innovations at Tesco was the implementation of the *continuous replenishment* concept in their distribution network. This is Tesco's equivalent of a just-in-time delivery system, which means products and orders flow through the supply chain quickly and frequently rather than waiting to be processed in batches. Each store receives up to seven deliveries per day, and the deliveries are organized to minimize the time and effort required for replenishment on the shop floor. This is achieved with a fleet of lorries that is second only to that of the Royal Mail in the United Kingdom.[24]

In the retail stores, the continuous replenishment system is rounded off with another of Tesco's innovations, known as the *one-touch replenishment* system, which is basically *shelves on wheels*. Tesco has nearly eliminated the need for store backrooms by putting stock on merchandising units that can be wheeled onto the shop floor or, in the case of shelf-ready packaging, be replenished very quickly with minimal effort in very little time. Other products are merchandised and displayed on wheelbase trolleys; once these trolleys are unloaded from the lorry, they are pulled from the landing bay and brought onto the shop floor. Products are, therefore, touched only once physically by the Tesco staff. This reduces significantly the amount of time for product handling and all related operational costs.[25] This also means less room for storage and more room for sales. In the spirit of continuous improvement, Tesco continues to invest in and upgrade their supply chain operations, in-store operations, and ordering systems to induce further effectiveness and efficiency.[26] The company streamlined its operations through their Tesco Digital program, which delivered a £550 million increase in profitability during 2009 alone.[27] This companywide initiative has also facilitated the minimization of stock holdings.

Tesco Clubcard program itself was turned "green" as part of the company's integrated corporate social responsibility (CSR) strategy and their 10-point Community Plan to increase greenness at their stores and improve their position within the community. The green Clubcard points reward customers who bring their own bags when shopping, with the aim to reduce the number of carrier bags handed out in stores and speed up the adoption of thicker and degradable bags. In the UK, Tesco gives out over 1 billion green Clubcard points, worth over £10 million per year to customers who recycle aluminum cans, printer cartridges, and mobile phones in Tesco stores, choose bagless home delivery, buy home insulation, or reuse carrier bags. In 2010, the green Clubcard points initiative was launched in most of Tesco's international stores.[28]

We define the final stage in Tesco's evolution in their recent history as the *lean and green* stage. This was marked by the introduction of Community (or sustainability) objectives in 2007, into the corporate strategy and integrating them into the Steering Wheel. As mentioned earlier, in January 2007, Sir Terry Leahy, then CEO of Tesco, delivered a landmark speech outlining the company's ambitious commitments to sustainability. Tesco has been drawing on their lean continuous improvement culture and their clear strengths in alignment and engagement to deliver their sustainability targets. Tesco's commitment to reduce CO_2 emissions throughout the entire supply chain—from suppliers to consumers—is evidenced by the fact that sustainability is directly quantified and measured in the Steering Wheel, by issuing annual Corporate Responsibility Reports since 2001, and by a great number of appearances, speeches, and articles released in the press by the company's two CEOs since 2007. Tesco's strategic environmental objectives are as follows:[29]

- Become a zero-carbon business by 2050
- Halve the distribution emissions of each case of goods delivered by 2012, against a baseline of 2006
- New stores built between 2007 and 2020 to emit half the CO_2 of a new store in 2006
- Reduce the emissions of the products in the supply chain by 30% by 2020
- Find ways to help customers reduce their own carbon footprints by 50% by 2020

In early 2010, Tesco announced a three-part approach to tackling climate change:[30]

- **Leading by example:** Tesco will keep cutting its own emissions as a business, working on energy efficiency and using new technology to produce renewable energy at its stores and depots.
- **Working with others:** Tesco will work even more closely with its suppliers to reduce emissions embedded in the products that they sell.
- **Empowering customers:** Tesco wants to help customers to lead low-carbon lives by making green products more affordable and more available, and by improving information through carbon labeling.

As a result of their activities, Tesco has, most notably, managed to decouple business growth from carbon emissions. As the net sales grew by 8.8% in 2010–2011, Tesco's carbon footprint increased by only 2%. They have also introduced carbon footprint labels on more than 500 items in stores, which in the UK account for £1.3 billion in sales per year.[31] Introduction of carbon labels has been a major step in the UK in educating shoppers about their own shopping basket footprint, while also providing clear targets for Tesco and suppliers to strive for in reducing environmental impacts. Table 6.3 provides a summary of some of Tesco's most important achievements in meeting their green strategy objectives up to 2011.

Conclusions

Charles Darwin famously said: "It is not the strongest of the species nor the most intelligent that survives. It is the one that is most adaptable to change." The same is true for businesses. As markets, consumers, and technologies change, businesses need to adapt their visions and strategies. But a bigger challenge is to ensure that the entire body of the organization—all individuals, functions, and processes—also adapt to the change. This is the significance of successful strategy deployment. Adopting lean and green operations depends on the organization's ability to adopt the new vision and their capability to cascade it throughout the entire company.

In this case study we saw how a leading retailer, Tesco, has deployed its lean and green vision and strategy across thousands of stores, hundreds of distribution centers, and nearly a half-million staff globally. Strategy deployment is about alignment and engagement; in Tesco, all operations, decisions, and projects are aligned to the core purpose, vision, values, and strategy of the organization, and the personnel are engaged in this effort. Tesco goes even beyond staff engagement and engages its customers through a series of innovations and initiatives that aim at gaining their loyalty. Tesco achieves both alignment and engagement by implementing tools such as the Steering Wheel in operations and the Clubcard in marketing. Tesco set clear objectives and targets and cascaded a whole raft of KPIs across different functions to ensure meeting their strategic objectives. By adding Community to their strategy and to their Steering Wheel, Tesco has given a clear voice for sustainability to be taken into account in day-to-day operations.

Tesco is one of the most aligned and engaged companies we have worked with. In this chapter we explained the evolution of lean and green thinking in Tesco. Tesco started its journey by looking at focused internal efficiency gains in the early 1990s. They soon adopted lean as a corporate strategy and also launched the Step Change initiative in 1999 to deliver an improved shopping experience and to run their operations better for customers, simpler for staff, and cheaper for Tesco.[32] Adoption of lean thinking led Tesco to realize opportunities in the end-to-end supply chain, such as one-touch replenishment and shelf-ready packaging solutions described in

Table 6.3   Tesco's Lean and Green Strategic Achievements in 2010–11

- Decoupled business growth from carbon emissions: net sales area grew by 8.8%, while carbon footprint increased by only 2%
- Reduced emissions from refrigeration (UK) by 165,000 tons of CO_2e by reducing leakage of fridge gas
- Tesco named as top retailer in the Carbon Disclosure Project's 2010 UK FTSE 350
- Reduced the absolute CO_2e emissions in the UK by 5%
- Opened zero-carbon stores in Czech Republic and Thailand
- Completed two more zero-carbon stores in the UK
- Completed the zero-carbon Tesco's Leadership Academy in South Korea
- Installed interactive energy boards in the UK and Ireland stores projected to save 15,000 tons of CO_2e, equating to a cost saving of £2 million every year
- Energy boards won Best Energy Reduction Project of the Year at the Energy Event 2010
- Reduced HFC emissions by 15% compared to 2009
- Pledged to begin phasing out HFC refrigerants from 2015
- Installed four wind turbines to power three UK depots using renewable energy; each turbine will produce enough energy to power 500 homes, saving around 3,200 tons of CO_2e
- UK household food and drink waste reduced by 4%

- Installed solar panels on tens of Tesco stores in Arizona, which will generate about 20% of the energy used in-store
- Reduced the maximum speed of UK vehicles to 50 mph, projected to reduce fuel consumption by up to 3%
- UK rail network saved 6 million road miles and over 8,000 tons of CO_2e
- Opened the first depot to be powered entirely by renewable energy in Widnes, UK
- Daventry depot has direct rail freight access for up to 8 trains per day, which is projected to save nearly 20,000 tons of CO_2e per year
- Network optimization has saved 12 million road miles in the UK, and over 16,000 tons of CO_2e
- In the UK, carbon-labeled products account for £1.3 billion of sales each year
- Since January 2008, 1,100 products in the UK have been carbon footprinted
- Since January 2008, 525 everyday products in the UK have been carbon labeled
- Since 2009, no waste from UK stores has gone directly to landfill
- In the UK, all cardboard, metal, office paper, plastic, chicken fat, and cooking oil are recycled
- Tesco is the largest retail recycler of cardboard in the UK, processing nearly 300,000 tons per year
- 228 tons of batteries recycled this year
- Product and packaging waste in the supply chain reduced by 5%

Source: Collated from Tesco's *CSR Annual Report* 2011.

this chapter. We call this stage of Tesco's lean and green evolution the *end-to-end efficiency* era.

Combining their Step Change program and lean supply chain management abilities with a deep understanding of consumer needs through the Clubcard and Tesco.com enabled Tesco to shift to a more advanced stage that we regard as the *end-to-end effectiveness* stage around 2000. We explained how the Clubcard is one of the sharpest tools in Tesco's repertoire, giving them a unique ability to target their promotions, discount vouchers, marketing campaigns, and even store ranging to exactly the needs of the individual shoppers. This is the ultimate application of the Genchi Genbutsu principle—getting really close to the shop floor and to their customers to earn their lifetime loyalty. Finally, we explained that the final stage in Tesco's evolution is the *lean and green* stage. Tesco adopted lean in the mid-1990s, but since Sir Terry Leahy's *carbon speech* in January 2007, the company has been putting sustainability at the heart of its vision, strategy, and operations. Tesco even introduced a new segment in the Steering Wheel, to ensure that the top and the bottom of the organization are aligned to this new strategy and that corporate responsibility messages are actualized beyond a set of nice words. In 2012 Tesco announced 5% year-to-year reduction in their total greenhouse gas emissions from 2011, this is a trend that has been growing for the past several years.

Tesco continues to prosper and grow profitably as it keeps evolving its strategy and operations. What is incredible about Tesco is its ability to deploy its strategic intents in very dynamic and efficient ways. Tesco staff is used to putting company purpose, vision, values, and strategy at the heart of their day-to-day operation.

Endnote

1. The contents of this case study are the authors' own opinions and do not necessarily reflect those of Tesco Plc. Information presented in this case study was gathered from various publicly available resources.
2. Data obtained from Tesco Corporate 3 reports.
3. Sir Terry Leahy, "Every Little Helps," speech delivered at the London Business Forum, October 2010,www.londonbusinessforum.com.
4. Liptrot, H. 2005. Tesco: Supermarket Superpower. BBC Money Programme, June 3. http://news.bbc.co.uk/2/hi/business/4605115.stm
5. "Our Values," Tesco PLC. 2012. http://www.tescoplc.com/index.asp?pageid=10
6. "Our Strategy," Tesco PLC, 2012, http://www.tescoplc.com/.

7. Sir Terry Leahy, "Every Little Helps," speech delivered at the London Business Forum, October 2010,www.londonbusinessforum.com.
8. J. Tippett, *Informing Choice, Leading Change: Towards a Sustainable Future* (Manchester, UK: University of Manchester Sustainable Consumption Institute, 2009), http://www.ourfutureplanet.org/newsletters/resources/Skills_in_the_workplace_SCI_Project_Report_July2009.pdf.
9. Tesco Corporate Responsibility Review 2005: Measuring Our Performance, Tesco PLC, 2005, www.tesco.com/csr/c/c2.html.
10. B. J. Witcher and V. S. Chau, "Contrasting Uses of Balanced Scorecards: Case Studies at Two UK Companies," *Strategic Change* 17 (2008): 101–114. Published online in Wiley InterScience (www.interscience.wiley.com) doi: 10.1002/jsc.819.
11. "Performance Management at TESCO," Scribd, http://www.scribd.com/doc/45964067/CMI-Assignment-7002-Strategic-Performance-Management, accessed April 4, 2012.
12. Sir Terry Leahy, 2007. "Tesco at Forum for the Future," Food Climate Research Network, January 2007, http://www.tesco.com/climatechange/speech.asp.
13. Tesco Corporate Responsibility Report: Community Promises, Tesco PLC, 2011. http://www.tescoplc.com/index.asp?pageid=93
14. D. Jones and P. Clarke, "Creating a Customer Driven Supply Chain," *ECR Journal 2*, no. 2 (2002). http://ecr-all.org/content/ecropedia_element.php?ID=12401
15. Tesco Annual Report, Tesco PLC, 2001.
16. Tesco Annual Report, Tesco PLC, 2010.
17. Rebecca Marston, "Tesco Triumph's under Sir Terry," *BBC News*, June 8, 2010, http://www.bbc.co.uk/news/10263953.
18. Sir Terry Leahy, "Every Little Helps," speech delivered at the London Business Forum, October 2010. www.londonbusinessforum.com.
19. Sean Poulter, "No Loyalty! Tesco Sells Details of Your Shopping Habits for £53M," *Daily Mail* online, March 17, 2011, http://www.dailymail.co.uk/news/article-1365512/Tesco-sells-details-shopping-habits-53m.html#ixzz1rb0nQAWd.
20. The Value range recently changed its name to Everday Value.
21. Jones and Clarke, 2002 "Creating a Customer Driven Supply Chain."
22. The probability of finding all 40 items is 0.985 to the power of 40, which is 54.6%.
23. Liptrot, H. 2005. Tesco: Supermarket Superpower. BBC Money Programme, June 3. http://news.bbc.co.uk/2/hi/business/4605115.stm
24. Ibid.
25. Jones and Clarke, "Creating a Customer Driven Supply Chain."
26. T. Abeysinghe, "Roaring Tigers, Rising Dragon," *Straits Times* (Singapore), February 2, 2010, http://www.fas.nus.edu.sg/ecs/scape/doc/ST-NUS-Econ-Series-02Feb10.pdf, accessed July 7, 2010.
27. INFORTEC, "Tesco Selects ORTEC to Optimize Its Supply Chain," ORTEC company magazine, 2010/11, http://www.ortec.nl/~/media/Files/Infortec/Nederlands/2010_INFORTEC_12_en.ashx.

28. Tesco Annual Report, Tesco PLC, 2010. Tesco PLC annual report and financial statements 2010: A business for a new decade. http://www.tescoplc.com/files/pdf/reports/annual_report_2010.pdf
29. Tesco Corporate Responsibility Report: Community Promises, Tesco PLC, 2011. http://www.tescoplc.com/index.asp?pageid=93
30. Ibid.
31. Ibid.
32. Tesco Annual Report and Review, Tesco PLC,, 1999.

Chapter 7

Lean and Green Supply Chain Collaboration

In this chapter we focus on the Supply Chain Management element of the Lean & Green Business System model (Figure 7.1) and will demonstrate the power of collaboration across the end-to-end process. This chapter tells the story of Marks and Spencer (M&S) and their remarkable success in implementing lean and green along the end-to-end supply chain. M&S is a leading green brand in the world of retailing that has extended their internal sustainability ambitions beyond their own organization embedding it across the whole process. You will also find a second case study belonging to MAS Intimates in Sri Lanka which is a leading manufacturer of lingerie for brands such as Victoria's Secret and M&S.

Lean and Green Supply Chain Management: Case Study of Marks & Spencer[1]

A Brief History

Today M&S is recognized as one of the leading retailers in the United Kingdom, with more than 21 million people visiting their 700 UK and expanding range of overseas stores each week. The retailer offers clothing and home products as well as fresh and prepared foods, each accounting for around 50% of the business and sourced from 2,000 suppliers all over the globe. In terms of market share, M&S is the UK's biggest clothing retailer,

Figure 7.1   Lean and Green Business System model.

holding 12.3% by volume, and making up 3.9% of the UK food market. The M&S in-store cafe is now the third largest coffee chain in the UK. As a whole, the company directly employed in the region of 78,000 people in 2010–2011, and their recorded sales were £9.7 billion with a healthy underlying group profit before tax of £714.3 million.

From its founding years, the mantra of *quality, value, service, innovation, and trust* has always been central, and current members of the M&S team feel that these have always informed the company's sense of corporate responsibility. Numerous members of M&S management feel that the company has moved through different stages of putting the core principle of trust into action: starting off with philanthropic actions, moving to a wider perspective of corporate social responsibility (CSR), and since 2007, to a much more strategic version of sustainability underpinned by a strong business rationale.

M&S has moved through a classical trajectory of increasing responsibility to a growing range of stakeholders as illustrated in Figure 7.2. Interestingly, this trajectory also traces a path through the stages of lean and green maturity toward a true symbiosis of the two concepts as discussed in Chapter 3 (from benign relationship to synergy to symbiosis of economic and environmental objectives).

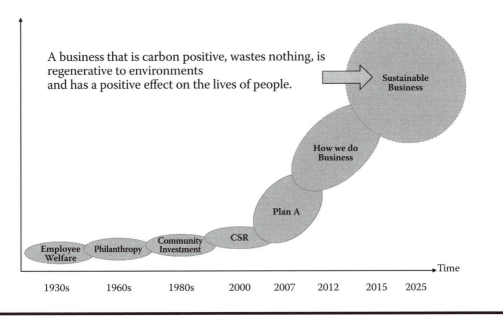

A business that is carbon positive, wastes nothing, is regenerative to environments and has a positive effect on the lives of people.

Sustainable Business

How we do Business

Plan A

CSR

Community Investment

Employee Welfare

Philanthropy

Time

1930s   1960s   1980s   2000   2007   2012   2015   2025

Figure 7.2   Towards Sustainability (Reproduced with permission from © Marks and Spencer, 2011.)

Picking up this story before 2007, Stuart Rose, CEO of M&S at the time, highlighted that despite doing a great deal of impressive work behind the scenes to ensure M&S was a responsible corporation, the company was not

CARMEL MCQUAID, CLIMATE CHANGE MANAGER, MARKS AND SPENCER

After completing an undergraduate degree in chemical engineering, Carmel started her working life with a graduate placement at a leading chemical company overseas. Chemical engineering is all about resource efficiency—creating the maximum value by clever conversion of precious raw material resources. So the philosophy of lean and green can be part of the day job for a chemical engineer. During her early *apprenticeship*, Carmel learned two important lessons—first, operations are much more about people than machines, and second, a business can only operate with permission from society. Business must provide goods that people want in a socially acceptable way to be profitable. In her first role, "concern for the environment was immediate and practical rather than theoretical, because in Germany, where I used to work, if you were the factory manager and something went wrong, you could personally go to jail. Our factory was located in the middle of a city with neighbors overlooking the factory. I ran the smelliest part of the factory, so I'd have to deal with the complaints. I had a real sense that we were only able to continue in business as long as we ran a safe, clean operation and this needed motivated and engaged workers who focused on adding maximum value to the product."

Later, however, Carmel moved into a supply chain role and worked in factories, distribution, and supply of products to customers. There, she was able to experience being at the back end of the supply chain and see the "bullwhip effect" firsthand. At times suppliers, without any advance communication, at the very last minute posted an order and brought down great pressure to maintain supply to prevent upstream factory shutdown or late delivery. Moving into consultancy, Carmel spent a number of years working on manufacturing and supply chains. It was during this stage that Carmel started to become involved with larger lean projects, initially on a project for the UK tax office and later for the Department of Work and Pensions. Here she quickly realized that while most people *got* the *tools and techniques* side of lean, it was a challenge to bring them to a place where they understood the real philosophy behind it and how to culturally make the change.

In 2007, Carmel moved to work for the Plan A team at M&S; her biggest satisfaction comes from building capability in her colleagues across the business so they realize that improving social and environmental impacts also drives business value through innovation and efficiency gains.

making the most of communicating this to the customer. Indeed, current employees recognize that people have always thought M&S to be "doing good things," but often found it difficult to explain why they held such belief. To help make customers more aware, M&S initiated the "Look Behind the Label" campaign, as part of a wider renovation of the public image, in which customers were provided with extra information about the work of the retailer.

The Look Behind the Label initiative was particularly pertinent because unlike most UK retailers, the majority of the goods that the company sells are its own brand. The initiative allowed M&S to directly inform customers about the intricate extent to which the company undertook food safety testing or ensured that dyes used in the production of its clothes were safe for both people and the environment. Overall, the campaign is considered to be one of the most successful that the organization has ever undertaken. Eighty percent of the customers said that they really do want to know more about how the products are made, the conditions in the factories, and the standards that are employed throughout the supply chain. It was this initiative that made the company really realize the extent to which the customers cared about the life story of food and clothes, if it was told in an engaging and meaningful way.

Around the same time, climate change was high on the minds of the general public and the media; and more generally while the Earth has little more to give, it is possible to change and learn to live within the Earth's means. Consumer research told M&S that their consumers "expected government and big business to take the lead on environmental issues, they felt a sense of disempowerment and were looking for leadership". Drawing on the confidence inspired by the Look behind the Label initiative, M&S decided to take the lead in acting on behalf of public opinion on the issue along with other key issues of waste, raw materials, health, and treating people fairly.

So in January 2007, M&S launched the Plan A program. In the words of the retailer's official website, "We're doing this because it's what you want us to do. It's also the right thing to do. We're calling it Plan A because we believe it's now the only way to do business."[2] It is Plan A because there is no Plan B for our planet. The program initially emerged with a total of 100 commitments to be achieved in 5 years. The aim was to make M&S carbon neutral by 2012 and send no waste to landfill. Plan A aimed to help customers and employees achieve a healthier lifestyle and to improve the lives of all those who worked in the company's supply chains with fair wages and improved labor conditions. The company also hoped to reduce its use of energy in all its stores by 25% per square foot of floor space, as well as achieve a 20% efficiency increase in all its offices and warehouses.

In contrast to some ethical business initiatives, which only operationalize change when there is a clear business case, M&S embarked upon their Plan A journey with the perspective of such objectives being *the right thing to do*. Indeed, when Plan A was launched, it was believed that it would cost £200 million over the 5-year time frame. However, as with other experiences talked about in this book, environmental and wider ethical initiatives have also been strongly supported by a business case. By 2009–2010, Plan A had generated £50 million and in 2010–2011 this was up to £70 million, which has been invested back into the business.

According to Carmel McQuaid, Climate Change Manager at Marks and Spencer, the rationale for Plan A is four fold: (1) brand enhancement and protection, (2) operational efficiency, (3) increasing staff motivation, and (4) driving innovation and improving resilience.[3] Due to the depth of this commitment, an updated version of Plan A was published in 2010 to increase the total number of commitments to 180, spread across seven pillars, all of which are to be completed by 2015. A sample of these commitments is reproduced in Table 7.1 and the seven pillars are illustrated in Figure 7.3.

The ultimate goal of M&S is to become the world's most sustainable major retailer. The first two pillars are (1) the involvement of customers and (2) making Plan A the way that M&S does business. The remaining five objectives cover (3) climate change, (4) waste, (5) sustainable raw materials, (6) fair partner, and (7) health. Members of the current team feel that the plan is underpinned by existing achievements and therefore is rigorous and credible, even though the inclusion of stretch goals left promises that at the beginning "quite frankly, no one knew exactly how they would get to."[4]

Engaging Staff Internally

One of the key aspects of Plan A is that it has sought to engage the entire spectrum of the workforce. Each of the 690 stores has a Plan A Champion who volunteers to take responsibility for promoting the plan among other staff members.[5] The company offers days where the champions can meet to share ideas and discuss various challenges that they might be facing. Others are asked to constantly engage with the issues and practices of Plan A through widespread communications, such as posters, the staff magazine, and emails, as well as team debriefs and the intranet site, which hosts a carbon footprint calculator, a car-sharing tool, and a blog for staff to ask questions and share ideas for improving environmental performance.[6] It is through these means that staff members are asked to play a part in the bigger program and are empowered to understand what contribution they can make by reducing the use of electricity and paper or increasing the amount of waste that can be recovered as new materials.

Another noteworthy feature of Plan A is that it has been embedded into the governance and accountability structures of M&S. At the highest level, the board of directors now has bimonthly meetings to ensure the executive team plays an active role in defining Plan A strategy. Executive directors at M&S, as well as members of the management committee, have Plan A targets that directly contribute to their personal performance bonus.

Table 7.1   Sample of Plan A Commitments

Climate Change	Waste	Sustainable Raw Materials	Fair Partner	Health
Aiming to make all our UK and Republic of Ireland operations (stores, offices, warehouses, business travel and logistics) carbon neutral.	Aiming to ensure that M&S operations in the UK and Republic of Ireland will send no waste to landfill.	Producing our fruit, vegetables, salads, and meat to independent environmental standards.	Extending our Milk Buying Pledge, which offers a guaranteed set price, to other types of food in consultation with our farmers.	Aiming to maintain our position of offering at least 30% healthier food lines.
Reducing the amount of energy we use in our stores by 25% per square foot of floor space.	Sending no waste to landfill from M&S store construction programs.	Maintaining our non-genetically modified (GM) food policy.	Ethical compliance monitoring.	Developing agricultural projects to provide naturally enriched foods.
Working in partnership to build and operate anaerobic digestion facilities to generate renewable electricity.	Having reduced food waste, we plan to work with suppliers to improve stock planning by developing the accuracy of systems for forecasting demand.	Phasing out pesticides that will be banned by the EU, in our fruit, vegetable, and salad production across the world, ahead of legislation.	Increasing our use of small and small local suppliers by improving the understanding of our buying teams of the different needs of small producers.	Ensuring all our fresh salmon meets our new Loch Muir standards (enriched with Omega-3) during 2007–2008.

(continued)

Table 7.1   Sample of Plan A Commitments (continued)

Climate Change	Waste	Sustainable Raw Materials	Fair Partner	Health
Having introduced new transport technologies, aim to conduct further trials and extend the usage of successful innovations.	Improving our use of recycled and recyclable materials in consumable items used in our stores and offices.	Ensuring all the wild fish we sell is Marine Stewardship Council–certified or, where MSC is not available, another equivalent independent standard.	Introducing an enhanced web-based system that further improves the information we have about factories and raw material suppliers within three years.	Making further reductions to the amount of salt in our foods by working to M&S targets, many of which go beyond those set by the Food Standards Agency.
Sourcing as much food as possible from the UK and Republic of Ireland.	Reducing the weight of nonglass packaging by 25%.	Reducing the environmental impact of the textiles we sell throughout our supply chain.	Further developing our responsible buying guidelines and integrating them into our buying processes.	Introducing 1,500 Healthy Eating Advisers in stores and extend this training to our food section employees by the start of 2010.

Source: Marks & Spencer, *How We Do Business: Reports,* http://corporate.marksand-spencer.com/howwedobusiness/hwdb_reports

Furthermore, M&S has gone to the extent of having the top 40 Plan A commitments externally audited and assured by an independent auditor on an annual basis.[7]

Naturally, all this requires a financial commitment, and for this reason M&S (2010) has placed £50 million (around $75 million) in a Plan A Innovation Fund to be spent over the next five years.[8] The Fund has already supported a range of central projects including trials on extending the life of food (which in turn reduces waste), but has also been made available for staff-instigated projects such as funding ways to reduce food waste, develop more sustainable fabrics, and improve energy and water efficiency resulting in considerable economic, brand value, and ethical benefits for M&S as a business and for the society at large.

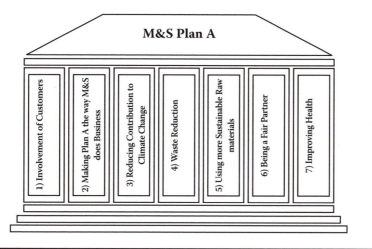

Figure 7.3   The Seven Pillars of Plan A (Marks & Spencer, *How We Do Business Reports*, **http://corporate.marksandspencer.com/howwedobusiness/hwdb_reports.)**

Engaging Suppliers

The first year of Plan A mostly involved communicating the manifesto for change to the M&S supply base. Indeed, the approach of M&S has been very much one of collaboration with their suppliers right from the very beginning. Thinking back, current members of the Plan A team recall that there were about 10 to 15% of the existing suppliers who were probably already taking action on some of the commitments, but had just not communicated this yet. These suppliers came forward with their own projects on how some of the targets could be met; for example, clothing manufacturers highlighted their eco-factories and food suppliers set about chasing the targets set for reduction of packaging. On the other hand, one senior executive remarked that some other suppliers were less enthusiastic: "The first year, back in 2007, people were watchful wondering if Plan A would be a passing fad."[9]

With this in mind, in the second year M&S started to engage more widely with their supply base to embed Plan A into their objectives and operations. Initially, the approach was to continue to encourage collaboration and consultation and draw on the technical knowledge and experience of suppliers, particularly as M&S acknowledges that it is not necessarily an expert on many of these operational issues.[10] In the early years, pilot projects played an important role in accelerating solutions and learning. For example, in setting up model eco-factories, opportunities were identified around improving the building and the building infrastructure, so M&S was able to take a lot of knowledge from its store team who had worked on eco-stores to work with clothing producers overseas and food producers in the UK.

Having completed a number of single-issue pilots, for example, model eco-factories and model ethical factories, M&S was keen to pilot best environmental and ethical practice within a single production facility and so a project at Flixton, UK, was set up incorporating environmental, ethical, and lean programs simultaneously.[11] More recently, M&S has held a series of high-profile Plan A supplier conferences in the UK, where only one event in February 2011 was attended by over 1,200 people, and around the world where their suppliers are based.

Key Achievements of Plan A

In 2011, M&S had already delivered 95 of the 180 Plan A commitments set in 2007. From the remaining 85 commitments, 77 were on target, one (regarding biodiesel) was on hold, and only seven were behind plan. By 2011, the retailer reported that gains from Plan A had contributed a net benefit of over £70 million, up from the £50 million in 2010. While the full extent of M&S achievements and programs are published on their website and in various progress reports, we cover some of the most prominent activities and the current level of achievement.[12]

Plan A as a Key Source of Innovation

Arguably, the biggest achievement of Plan A is in *its role as a source of innovation and inspiration across the business and across its supply chain.* Members of staff and suppliers alike engage in initiatives funded or otherwise encouraged by Plan A delivering substantial benefits to the businesses, the consumers, and the society. For example, not only has M&S reduced its own energy consumption, but it has encouraged similar improvements by its suppliers as well as assisting both workers and customers to implement such changes in their own lives. In this chapter, we also discuss achievement of suppliers such as MAS Intimates.

Richard Gillies, director of Plan A, regards the program as a prime source of innovation realizing economic prosperity and business longevity for M&S, while supporting a more sustainable future for the shoppers and the society at large, "a source of innovation that no one including M&S can afford to miss."[13]

Different companies use different initiatives to confront the perils of *status quo* and to realign themselves with the ever-evolving needs of their customers. Some prefer to use terms such as *lean* and *lean-sigma*, others simply call it continuous improvement or the company's way. Plan A is Marks and Spencer's drive for betterment and their critical commitment to realizing tomorrow better than today.

Plan A and Enhancement of the M&S Brand

After joining Marks and Spencer in May 2010, the new CEO Marc Bolland, a brand marketer himself, recognized that Plan A is "a strong point of difference" for shoppers who are faced with a plethora of retail brands competing for a share of their disposable income. He made a strong point saying, "I will carry [Plan A] forward as a big identifier for the brand. We should bring it out more. We should love it more."[14] Being a responsible retailer is a key part of Marks and Spencer's identity and core to their brand value. Bolland made no hesitations to recognize this factor when he said, "Sustainability is central to how we do business and Plan A will help us to stay ahead in a fast moving world."[15]

It is hardly surprising that Marks and Spencer is often cited as one of the leading responsible brands worldwide. For example, a report produced by Brand Sustainable Futures in 2010 put M&S at the top of 150 global retailers[13] in spite of the fact that their global presence remains limited to a few countries with a strong focus on the United Kingdom market. In the following we provide a summary of some of the more tangible aspects and achievements of Plan A to date while emphasizing that the real benefits of Plan A, both in terms of brand value and becoming a source of continuous innovation within the business and its supply chain, are extremely difficult to quantify.

Plan A, Resource Efficiency, and Waste Reduction

One of the most pleasing achievements for M&S is to have cut their carbon emissions by 13%, which is a reduction of over 90,000 tons CO_2 equivalent between 2006–2007 and 2011, despite the fact that the sales floor footage has continued to grow. This means that the company is on target to become carbon neutral, although it is behind its plans in identifying steps to meet water efficiency targets.[14] In terms of energy usage and energy intensity, by 2008 M&S had three wind turbines generating enough energy to power three stores, and in 2009, began to purchase the remainder of its energy needs from Npower's renewable program. In total, M&S by 2011 supplied around 60% of its electricity from renewable sources.[16] Another significant reduction in carbon footprint output from the retailer's activities came from reducing the number of business flights taken by employees by 14% per person between 2007–2008 and 2011. In order to use the energy that is drawn in the most effective way, the company has improved in-store energy efficiency by 23% (after weather adjustments) and warehouses by 24% against 2006–2007 levels.[17]

M&S has also improved the fuel efficiency of their general merchandise and food delivery fleets by 20% and 22%, respectively. While food warehouses have reduced water usage by 62% per square foot since 2007–2008, progress in stores and offices has been more difficult, though the company hopes that new initiatives will get them back on course to achieve 20% reduction by 2012.[18]

Seeking not only to change its own behavior, but to help other stakeholders shift their practices, M&S distributed 38,000 free home energy monitors and provided 4,000 homes with free insulation. Also, the company sells only large electrical appliances rated at A or above and has also started to assess and improve the efficiency of small electrical products using the Energy Star scheme and the European Union (EU) Eco-label.

As a flagship program, M&S continued to build on learning from ecostores and opened its first Sustainable Learning Store in Sheffield in April 2011. The building uses 100% LED (light-emitting diode) lighting, incorporates a green roof and living walls to support local biodiversity, and is the world's first fully Forest Stewardship Council–certified retail store to use 100% Forest Stewardship Council (FSC) timber. In recognition of these features, the building has been rated as "Excellent" under the Building Research Establishment Environmental Assessment Method (BREEAM).[19]

The retailer has also been working with the Waste Resources Action Programme (WRAP) in the United Kingdom, implementing their Sustainability Framework to try to reduce the use of materials in the first instance and maximize the use of those that can ultimately be recycled. Of the waste that is generated by the company, 94% was recycled in 2010–2011, including the remaining unsold food, 39% of which went to anaerobic digestion to generate energy. To reduce waste, M&S has been improving the forecasting and planning systems used to estimate sales volumes and place orders with suppliers. Indeed, this has been part of the benefits generated in the supply chain, as will be discussed in more detail later in this chapter.

The retailer also began a "marking down" practice, which means reducing the price of food items with very short shelf life left in order to sell them rather than wasting, as a result of which they have so far managed to reduce unsold food by about 37% by sales value.[20]

Nonglass packaging has been reduced by, on average, 26% per item compared with 2006–2007 and the amount of packaging used in the general merchandise home delivery business has been cut by 16% per parcel against 2008–2009. Specifically, the company has reduced clothing packaging by 36% and carrier bag usage by over 80%, principally by introducing a 5 pence charge across all food stores in the United Kingdom and Northern Ireland in 2008.[21] Introducing a charge for carrier bags also generates profits, which are donated to Groundwork, a charity that has provided over 120 parks, garden, and play areas across the United Kingdom.

In terms of the end of use for clothing items, one of the company's objectives has been to create partnerships to increase recycling. The retailer donates its own "seconds" to Oxfam and the Newlife foundation for disabled children, and encourages customers to give unwanted garments to Oxfam by allowing the charity to distribute vouchers for new clothes in exchange for donated items. Following the collection of some 152 million coat hangers through drop boxes in store, M&S has recycled 79% and donated 57 pence to UNICEF UK per box, an initiative that has generated at least £650,000 per year for three years.[22]

Plan A in the Extended Supply Chain

Marks and Spencer has been pushing lean and green initiatives into their supply chains in both food and general merchandise. In food, in order to uphold their commitment to sourcing as much as possible from the United Kingdom and the Republic of Ireland, the company continues its tradition of innovation by actively exploring ways to grow new crops in the UK, such as sugar snap peas, sweet potatoes, and melons, and to extend the UK growing seasons for fruit and vegetables in order to reduce carbon associated with air freight. For example, working with farmers, the company has extended the usual period for growing asparagus, now April to September, reducing the need for imports.[23]

In general merchandise, by working with the MAS Holdings Group, a garments manufacturer in Sri Lanka that produces lingerie for international brands, M&S has created the first-ever carbon-neutral bra. A bra can be a remarkably complex product, into the making of which goes a range of elastics, underwire, hooks, cups, laces, yarn, and embroidery. In total, it could add up to more than 20 components, and several packing materials, each one of which is typically sourced from a number of different suppliers. In order to create the carbon-neutral bra, MAS and M&S calculated the carbon footprint taking into account each item's entire life cycle from component manufacture to transportation and even customer usage. M&S invested $400,000 into designing and building a model eco-factory, which opened in 2008, and benefits from extraordinary environmentally friendly features such as large-scale solar panels for power generation, green roofs for insulation and temperature regulation, and energy-efficient lighting.[24] According to Vidhura Ralapanawe, the manager of sustainability and communications at MAS, the design for this extraordinary building evolved out of "an emphasis on the human factor."[25] The net result is better working conditions and better livelihood for several hundred Sri Lankan workers, while the lingerie is produced at no additional cost because the factory has applied the principles of lean thinking. For example, the dock-to-dock lead times for lingerie have been reduced from on average 56 days to 23 days.[26] But even more importantly, the factory is organized into self-managed teams where quality and productivity are organized on the line and a robust culture of continuous improvement generates thousands of ideas from the shop floor, dramatically reducing material wastes and significantly improving quality (rework) and productivity. Vidhura says, "We designed the factory to be a center for learning and inspiration."[27] Vidhura continues, "We turned our associates from machine operators into team members and encouraged the pride of workmanship." Since then, MAS has gone a step further to produce a carbon-neutral lingerie range, thus also engaging their raw material suppliers in the journey to lean and green.

By investing in this project and two other model eco-factories, the Plan A team was able to identify the most critical technologies and evaluate the business case for future investments. As a result of this and similar pilots, they identified three areas that had a payback period of less than three years for future installations, that is, lighting, heating control, and insulation, which then became a requirement for all top 100 supplier factories to implement and resulted in 10% reduction in their energy intensity.[28]

Like many organizations leading on their lean and green journey, M&S has made use of third-party governance and certification systems to increase the sustainability of many of its products in its supply chain. For example, purchases of Green Palm certificates for all M&S products containing palm oil is ongoing, and a further 35 food lines made with certified sustainable palm oil have been introduced.[29] Similarly, much of the canned beef and leather is also supplied from sustainable sources. Moreover, while the company has had a sustainable fishing policy for over a decade, it has more recently become the UK's first company to sign up for the World Wildlife Federation's Sustainble Seafood Charter.[30] At the same time, M&S has increased the use of recycled polyester in general merchandise products by 73% to over 1,900 tons (equivalent to 47 million two-liter plastic bottles), in order to reduce their resource intensity in the entire supply chain, one use of which has been to make over 300 million clothing care labels, all of which carry the instruction "Recycle with Oxfam."[31]

As M&S started to drive sustainability criteria into product supply chains, it became clear that a lean approach to supply chain management would also enable the achievement of sustainability goals. In fact, the Plan A Innovation Fund enabled introduction of end-to-end lean supply chain solutions as a pilot, which later led to further integration of lean into the supply chain. We will present a case study of lean supply chain management at M&S in the following.

To drive ethical improvements in the supply chain, M&S is driving progress in best practice with 15 ethical model factories now established across Bangladesh, Sri Lanka, and India, where 37,000 workers in their supply chains have received employment rights and health and safety training. A key element of the ethical model factories was working with employees to improve productivity, which allowed the sites to invest in worker training.

People Factor

M&S already recognized that employee engagement was fundamental to driving the environmental and labor standard objectives that had been committed to under Plan A. Improving energy efficiency or reducing the waste sent to landfill required employee engagement and their contribution was essential to reach the targets across stores, the head office, and the suppliers. According to Carmel McQuaid, if the company had decided to implement initiatives such as charging for carrier bags simply as a cost-saving initiative, "it would have been so much more difficult to implement this change in pricing, which could mean members of staff having difficult conversations with customers at the till point. But the very fact that carrier bag charging was about us trying to reduce waste and the impact on marine life through plastic disposal, that was a much more motivating factor for our employees and so they felt happier having the conversation with any customers who challenged the charging."[32]

LOUISE NICHOLLS, HEAD OF RESPONSIBLE SOURCING, MARKS AND SPENCER

Louise Nicholls has worked for M&S for more than twenty years. Louise first entered the company as a food technologist and after a short career break returned to work in Human Resources and then IT. Primarily, however, Louise has been responsible for developing and managing an M&S industry-leading ethical trade program. In 2007 Louise was part of the small M&S team who developed Plan A, and her team is responsible for supporting the strategic development and delivery of the M&S foods sustainability program. She deals with issues as diverse as sustainable fish sourcing, labor, lean manufacturing, environmental standards in factories, animal welfare, food miles, genetic modification, fair trade, and climate change. Collaboration, Louise believes, is key to driving change at scale and so she has held a number of external roles to encourage closer alignment of approaches. Louise was a board member on the Ethical Trading Initiative board from 2004 to 2007 and was instrumental in the development of Sedex, a global database for supply chains to share ethical and environmental data; she has sat on the Sedex board since its inception. Louise also chairs the Sedex Associate Auditor Group and is a board member of Global Social compliance program. Louise's team won the *Business in the Community Responsible Supply Chain* award for their M&S sustainability framework, which is a building-block approach to embedding ethical, environmental, and lean thinking in the supply chain.

Louise's team founded M&S supplier exchange, which brings together HR, environmental, and lean representatives from across the supply base to focus on common challenges, develop practical tools, work with experts, and attend "seeing is believing" visits. "Often the answer is in our supply base. I see M&S's role as a catalyst to enable faster uptake of the good practice across the supply chain and to recognize those suppliers who are demonstrating real leadership in their communities and sectors." A key part of her role is interaction with a wide range of stakeholders and suppliers. In the last twelve months she has traveled to India, America, Morocco, Kenya, and Europe, meeting with over 800 suppliers and local stakeholders to understand their issues and perspectives.

In general terms, the M&S staff feels that what people want, wherever they work, is to know "what the company vision is, and what part they play within it. If M&S workers see sustainability is a primary element of how we do business, if we don't want to generate waste, if we want to produce the product right first time, and if they have got the right training to understand the part that they play in, then they will set themselves up to do that, because they know that they are going to be measured on that."[33] The retailer is also proud of the fact that in 2011, 95% of employees took part in their annual "Our Say Survey," in which the overall positive score for M&S as an employer increased by one percentage point to 76%.

The Plan A team has been instrumental in wider adoption of lean thinking across M&S. It was the emphasis placed by lean on people engagement and bottom-up leadership that made it attractive to Plan A team members such as Carmel and Louise. For the Plan A team, one of the key learning experiences has been the focus on people engagement. This is interesting because as we explained in the previous chapter, "real lean" is also about respect for people as opposed to blindfolded application of the tools and techniques. The CEO of one clothing supplier factory commented to Richard Gillies, director of Plan A: "After years of seriously investing in lean thinking and training we had only ever realized about 70% of the benefit I knew existed. Once we focused on a green factory, then the engagement really kicked in. People were more motivated to do things that directly help their family and community rather than just making a company more profitable."[34]

Synergies between Plan A and Lean Thinking in M&S

Independently of Plan A, M&S had already recognized the value of lean thinking for some time and there were various lean initiatives, launched in areas such as new product development and food supply chain management, which incorporated both the principles and the tools of lean thinking. As part of an effort to develop a practical understanding of how to drive a combination of environmental, ethical, and commercial performance, M&S

launched the Model Factory program at a supplier plant in Flixton. This was the first factory to include all three elements rather than the previous single-issue focus under model eco-factories or model ethical factories.

To ensure access to the cutting edge of knowledge, M&S set up an external advisory group. In order to integrate the existing lean initiatives, M&S invited Professor Peter Hines to join the advisory group. As co-founder of the Lean Enterprise Research Centre at Cardiff University, Peter played an important role in influencing the wider adoption of lean thinking. The learning from this Model Factory program, prompted M&S to include lean as a formal element in their Plan A supplier scorecard, which we will highlight in the following.[35]

According to Louise Nicholls, head of Responsible Sourcing at Marks and Spencer,

> The more we visited our suppliers' factories and listened to suppliers' comments, the more we realized that 50% of a site's environmental performance is driven by trained and engaged people. So to drive environmental performance sites had a better business case to invest in raising their ethical standards, in having a more permanent well-trained workforce with good management and shop floor engagement. We see lean thinking, as using resources efficiently and focusing on processes that add value to the customer and as a facilitator to encourage end-to-end supply chain thinking with really efficient workplaces where workers are highly engaged and understand their place in driving forward a really successful business. So, for us, lean is the enabler for driving our environmental and ethical performance.[36]

The subsequent expansion and integration of lean principles within M&S operations was greatly assisted through the natural synergies it held with Plan A objectives.

Scorecard Mechanism to Embed Sustainability across the Supply Chain

In order to systematically drive the *triple bottom-line* goals of sustainability through their food supply chain, M&S began developing a Food Supplier Sustainability Framework. This provided better clarity to suppliers and enabled a fact-based discussion between buying teams and suppliers on environmental, ethical, and economic dimensions. Accordingly, three separate scorecards were developed for each one of these dimensions where lean thinking informed the economic framework (aka the lean scorecard).

The simultaneous use of lean, green, and ethical scorecards is the ultimate manifestation of Marks and Spencer's lean and green vision. However, the scorecard is a means to an end rather than an end itself. Suppliers fill in the scorecard rigorously as a way to diagnose their requirements for improvement and develop a roadmap for their journey. There have been multiple rounds of discussions, trials, and reviews, and although the process continues in some areas, all those involved agree that a sensible, pragmatic, and workable system has been devised. The use of the scorecard system has now been embedded across the food commercial operation. Giving the sustainability scorecard an equal weight as the commercial, technical, product development, and logistic scorecards has enabled food suppliers to engage the M&S food buying teams in sharing their sustainability agendas and priorities.

The introduction of key performance indictors (KPIs) for the M&S buying teams to increase the number of products they buy with an M&S sustainability attribute, of which achieving silver on the sustainability scorecard as an attribute has further helped to drive progress and embed sustainability in how M&S does business.[37]

According to a current member of the Plan A team, they spent a few months trying to come up with a common framework that both food and general merchandise would use, but it soon became obvious that this was not the right approach. At the moment, while the general merchandise unit does not follow a formal lean scorecard due to the nature of the relationships with the suppliers, there is a lot of lean inherent in the quality, productivity, and employee engagement improvements that underpins the ethical works carried out. Needless to mention that implementing sustainability initiatives across 2.7 billion individual items sourced from tens of thousands of factories, farms, and raw material sources is a continuous challenge for Marks and Spencer. A challenge that people like Louise and Carmel regard as *opportunity for improvement* and therefore volunteer to confront head on.

With this in mind, one of the benefits of Plan A is that it has unified the business and has given staff a language and a platform to start creating synergies. Indeed, Carmel feels that both Plan A and lean give the workforce a powerful tool, a shared vocabulary, and the understanding that accompanies it. Moreover, by using these approaches, Carmel thinks that "you give permission to people of all levels to challenge behavior at all levels, whereas they might not have been able to do so before. For example, we hear people challenge their colleagues excessive printing or air travel plans with a simple remark such as 'that's not very Plan A; is it?' It's ok now. Whereas before you would never turn to your colleague and say, 'Wow, you're using a lot of paper there aren't you.' At least you could never do it so constructively."[38] Carmel remarked that she has observed the same constructive challenge with lean as well on issues as simple as use of better IT solutions to minimize duplication of effort when editing shared documents. It is largely this foundation that has allowed Plan A to run across different departments.

Current Challenges

Despite having made a lot of progress since Plan A was launched in 2007, the business recognizes that they are only on a journey toward true sustainability. Improving the environmental performance of global supply chains is an incredibly complex challenge that any individual company will find incredibly difficult. A much wider issue raised by some[39] is that while improvements have been made in some parts of M&S's extended supply chain, they remain part of a wider system that has many unresolved sustainability issues. Nevertheless, the greening of individual links in supply chains is a step in the right direction. The application of lean and green thinking to a full value chain is an even bigger leap forward; something that we will discuss in more detail through an example below.

Lean and Green Intervention to Improve the M&S Supply Chain

In addition to developing the Food Supplier Sustainability Framework (the scorecard approach) and deploying it in improving the lean and green performance of its supply chain, Marks and Spencer has been carrying out more in-depth interventions in order to lead the change both internally in their own in-house operations and externally in their supplier base.

Generally speaking, there are considerable economic and environmental opportunities to be had across the whole Fast Moving Consumer Goods (FMCG) industry, particularly in the supply chain and on the boundaries between companies. For example, in 2003 the Food Chain Centre (FCC) was established in the UK to eliminate waste and pilot lean thinking in the end-to-end FMCG chains; the center commissioned examination of 33 food chains from farm to fork and later reported that on average, 20% of the costs in the food chain add no value and could be reduced or eliminated altogether.[40] Maybe more interestingly, the center stated that it is rare to find a shared understanding of consumer needs throughout the supply chain but when achieved, it helps greatly to promote trust and innovation.[41] This presents a huge opportunity for all stakeholders along FMCG chains, especially retailers, to improve their economic and environmental performance simultaneously.

The Plan A team in Marks and Spencer spotted this opportunity[42] and provided a seed corn innovation fund to pilot a study for delivering lean and green improvements in a selected M&S end-to-end food supply chain. Interestingly, Marks and Spencer is one of few companies that we know that has introduced lean thinking to their operations and into their extended supply chain through the green root. For the Plan A team, this was born out of a profound conviction that applying lean thinking methods to the extended supply chain also leads to the betterment of the environmental and ethical performance of the whole system.

Subsequently, M&S recruited a key supplier to introduce the first of a series of in-depth lean and green supply chain improvement projects. One of us led this intervention, which also involved many stakeholders across the whole chain working as a central supply chain improvement and redesign team. The first step in implementing the lean and green improvement project was to select a suitable supplier and a single suitable product line, often referred to as a stock-keeping unit (SKU) in retail terminology. It was important to identify a representative product, one that touched a number of suppliers, contained enough common ingredients and processes to allow for widespread learning, and one that would cover most of the environmental and social issues. Having identified a *ready-meal* product line for improvement, a team of stakeholders was put together from across the supply chain including senior managers from M&S, the ready-meal manufacturer, the packaging supplier, and the meat supplier to the ready-meal manufacturer.

This *lean and green improvement team*, which also included Plan A members, agreed to collectively map the selected product—a ready-meal—all the way from primary ingredients to the retail shelf following it through various manufacturing stages, warehousing, distribution, and retail operation.

The whole team committed to walk and map the entire chain collectively. The project was conducted over approximately three months involving ten action learning workshops where the whole team came together to map the current state of the supply chain or to design the future state and the action plan that followed. The intervention methodology was designed to take this senior team of stakeholders through a normative process of learning,[43] reflection, and planning for the future. This is fundamentally different from assigning a team of outsiders, for example, a group of external consultants, on a fact-finding mission and then trying to implement improvements by means of business case justifications.

The action learning intervention adopted by Plan A allowed a far more profound understanding of the current state and created a much greater degree of commitment to change by having the senior managers from different functions all involved in the process of diagnosis. The intervention involved various steps as outlined in Table 7.2.

The lean and green action learning improvement team walked the end-to-end process and mapped each stage in detail identifying every step on the shop floor as well as mapping the information flows that drive the physical flows of the product. Figure 7.4 is a simplified schematic of the current state of the supply chain from raw meat to the retail shelf showing both physical and information flows.

Even in this simplified map, it can be seen that the flow of information is more complex than the physical flow of the goods. The real complexity is several degrees more. Having mapped the end-to-end chain and applied the lean principles to look for waste and inefficiencies, the improvement team identified several key opportunities for improvement. The key opportunities can be listed under four headings:

1. Realigning product specification and the recipe to the consumer needs
2. Minimizing product in-store waste, enhancing shelf life, and increasing on-shelf availability (OSA)
3. Reducing inventory, reducing lead time, and reducing the number of product touches
4. Reducing demand amplification and demand distortion

In the following, we will explain each one of these opportunities in more detail.

Table 7.2 Stages of the Lean and Green Supply Chain Improvement Project

Workshop	Stage	Content
1	Initial workshop	• Team building and familiarization with the lean and supply chain improvement approaches • Constructing a big picture map of the entire chain
2–6	Detailed mapping of various firms along the chain	• Mapping of both physical processes and information flows within each firm • Mapping the current state from different perspectives
7	Constructing the current state map workshop	• Pulling together the information gathered from the previous sessions in order to create an overall current state map • Identifing key opportunities for improvement
8–9	Creating future state map workshop	• Understanding the cost and benefit of the key opportunities • Developing a future state map • Creating an action plan or roadmap for improvement
10	Executive Report	• Engagement with wider stakeholders • Feeding back the detailed action plan • Getting buy-in for implementation of the roadmap

Realigning Product Specification and Recipe to Consumer Needs

The team analyzed the waste levels along the chain and identified that there were *two key areas* where waste accrued along the chain: in retail stores and in primary manufacturing of steaks. Figure 7.5 is a representation of the waste along the chain calculated in terms of the real cost of production at each stage.

Notably, an estimated 40% of the total waste was in primary production of meat, in the cut and pack operations, most of which was in producing steaks that were overweight as compared to the specification, or in other

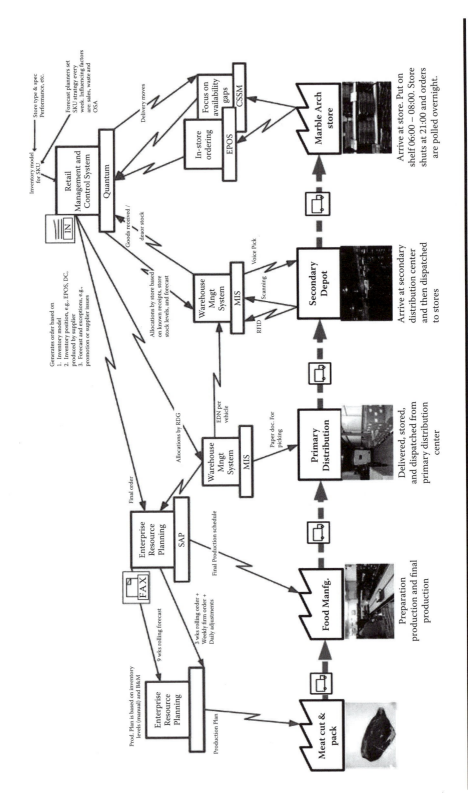

Figure 7.4   Current state map of the current state from butchery to retail shelf.

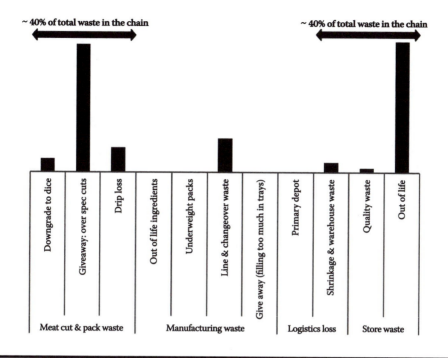

Figure 7.5   Analysis showing waste throughout the chain.

words in raw material giveaway. When the team walked the butchery operation, they noticed that the relevant cutting operation was manual, leading to
an assumption that the manual operation could be the source of the giveaway and that automation could potentially eliminate the concern.

However, following the principle of fact-based management, the supply chain improvement team decided to carry out a more thorough analysis
before applying a solution. Subsequently, rigorous sampling and measurement of the cut steaks was initiated, leading to the discovery that the manual cut operation was both capable and stable. Statistical analysis revealed
that the operator was reliably and predictably meeting the given specification ruling out the assumption that the manual operation was the cause of
the "giveaway" waste.

So where was the problem? The team later discovered that the relatively
high giveaway was due to a misinterpretation of the product specification,
where the specification was meant to be related to postmarinating weight
rather than premarinading weight.

In fact, this type of product specification misalignment or recipe misalignment occurs in many FMCG chains. One of us has worked with tens of
supply chains observing very similar opportunities; our research shows that

on average in 85% of the retail and food service chains there are significant specification alignment opportunities.[44] In this particular instance, the savings were not only financially significant, but also environmentally, since meat has a disproportionately large CO_2 footprint. *It was estimated that some 350 tons of $CO_{2\text{-}eq.}$ were saved, equal to driving around 2.5 million kilometers in an average-size automobile. This is equal to surfacing the earth at the equator around 50 times.*

Minimizing Product In-Store Waste and Increasing On-Shelf Availability

In the previous example, waste accrued in the upstream steak manufacturer, but the root causes were further downstream, resulting from a lack of communication about the meaning of the specification. In order to indentify and eradicate the problem at its root cause, the entire supply chain needed to collaborate closely and realign their activities with the actual needs of the shoppers.

This is often the case in supply chain management where waste in one area is rooted in a totally different part of the chain. We refer back to Figure 7.5, but this time we will focus on the waste that is recorded against the "last 100 yards." This is a term referring to the very final step in retail supply chains, from the backroom of the retail store to the shelf. Figure 7.5 shows that some 40% of the total waste in the chain is in the last 100 yards, where the biggest portion of that is categorized under products expiring or going out of life. Again, the waste in the last 100 yards can be attributed to causes that are further up the stream, for example, working with suppliers to minimize out-of-life waste by enhancing shelf life.

Typically, retailers have significant economic and environmental opportunities in minimizing the last 100 yards waste. In certain categories and for some SKUs the out-of-life waste is larger than the rest, for example, meat, dairy, and horticultural items. In-store waste is also a problem in terms of on-shelf availability. Most retailers know that when they try to increase on-shelf availability, wastage increases and vice versa, simply because more items on the shelf mean less likelihood to sell them in time to avoid out-of-life situations.

But this does not have to be the case all the time. Marks and Spencer is investing in supply chain systems that allow more frequent replenishment of the items on the shelf to ensure freshness, increased on-shelf availability, and reduced waste levels. Also they are working in their distribution process to enhance speed and accuracy of deliveries. *In this particular case, the improvement team identified ways, by changing the logistics cycles, to reduce the time the product spent in the pipeline by 50%.* The team mapped the physical time needed to take a single product from the factory to M&S stores and then asked, "Why should it take any longer to distribute the product in the logistics channel and how can we systematically eliminate those sources of delay?" For example, the product was idle for around 24 hours in a distribution center, which also meant energy consumed for chilling (part of which was in refrigerated lorry containers) as well as the loss of valuable shelf life.

In order to reduce waste in the downstream (in the last 100 yards), M&S needed to act on the sources of shelf life enhancement, which were further upstream, for example, in more efficient distribution. Similarly, the team identified that up to 20% of the total product shelf life could be consumed at the ready-meal manufacturer, even prior to dispatch. This was partly due to the need to cook the meat to forecast, putting it into chillers, and then assembling ready-meal packs according to the actual order. This is common practice in most food manufacturing firms, yet it means consuming shelf life, which is so critically needed in the last 100 yards to minimize the out-of-life waste levels. Therefore, the team sought new ways of working to allow better alignment between forecast and real orders. Altogether, the team identified solutions to reduce waste in the last 100 yards by about 30%, leading to significant potential economic and environmental savings.

IMPROVING WATER EFFICIENCY IN FOOD MANUFACTURING

In the process of cooking the steaks at the ready-meal manufacturer, large walk-in ovens are used and the supplier has to cool down the cooking racks before wheeling them out of the oven after each cooking cycle, due to both product quality and safety considerations. To achieve this, the interior of the ovens were sprayed with ambient temperature water for a period of 3.5 hours after 30 minutes of cooking. As it happened, the local operator had long wanted to save the water used for cooling the oven and quenching the product, which used to go into the drains. When the lean team started mapping the process and engaging with the staff, the team drew upon the knowledge of the local operator and the facility's environmental officers who suggested a whole raft of ideas for improving the water consumption. For example, and as a result of this interaction, it was suggested that if the water was chilled, the processing time could be cut by a full 3 hours, therefore taking only 30 minutes and consuming a significantly smaller amount of water. It was also suggested to replace nozzle heads with more effective ones for a small investment. The team estimated that annualized savings of up to 20 Olympic-size swimming pools could be achieved in this one process alone. Yet again, the importance of engaging with frontline staff was highlighted as a fundamental cultural change.

Reducing Inventory, Lead Time, and Number of Product Touches

A good measure for leanness is the amount of inventory a company carries. Lowering the inventory level obviously means reducing the product lead time, which in turn means a quicker cash-to-cash cycle—a very critical economic measure for any company. It is famously said that businesses are more likely to suffer due to cash cycle problems rather than profitability issues. The list of the companies with a healthy cash-to-cash cycle reads like the Who's Who of lean companies—Toyota, Dell, Tesco, and many more. In the previous section we explained why reducing inventory improves the overall performance of the chain economically and environmentally, and how the team targeted to reduce waste by reducing the lead time the product spends in the transportation pipeline before hitting the retail floor.

Put simply, reducing lead time in perishable product chains means less waste, especially in the last 100 yards of the supply chain where the impact of shorter life is more visible. More inventory also means more warehousing costs, for example, energy to chill depots as well as energy costs to heat adjacent spaces. Another reason for reducing lead time is that physical handling of products through the supply chain sometimes leads to damage and waste. This is generally assessed by the number of touches that are required by humans as the product moves down the supply chain. Some touches are essential and value adding, for example, those that involve cutting, marinating, and cooking the meat. Other touches are a result of the way the supply chain is designed, for example, moving the product from one place to another, replenishing shelves from a trolley, or loading the product onto the lorry. Where there is a physical *touch* there is always opportunity for waste.

As the team mapped process activities and time spent on the selected SKU, it was revealed that the product was touched 43 times between ready-meal manufacturing and the shelf, out of which only 7 touches were value adding in the best case. The product traveled through more than 100 steps in total, but not all the steps involved a physical touch by an employee. As a rule of thumb, and based on the authors' previous experiences in retail chains, the team targeted two thirds of the touches in the chain to be eliminated, that is, from 43 down to only 14 touches.

In recognition of the impact of the excessive touches, many supply chains now deploy some sort of one-touch replenishment unit, for example, shelves on wheels or shelf-ready packaging. The idea is to effortlessly move products on fixtures from factory floor to lorries, warehouses, retail stores, and even right out onto the shop floor. The advantage of this logistical arrangement is that the number of times individual units of product have to be handled is dramatically reduced. But more importantly, such solutions often lead to quicker replenishment logistics, reducing lead time and inventory levels in the supply chain. More usage of the quick replenishment solutions was thus another recommendation of the project.

Reducing Demand Amplification and Demand Distortion

Demand amplification is a well-known supply chain phenomenon prevalent in most supply chains, but it greatly affects the FMCG industry. Demand amplification refers to the effect whereby relatively small fluctuations in real consumer demand become progressively amplified when orders are passed upstream. There is a tendency in any multistage process, due to systems-driven reasons, for replenishment orders received by each upstream process to be more erratic than real requirements or demand at the next downstream process. This phenomenon, variably referred to as the bullwhip effect (due to the bullwhip shape when a graph is drawn) or demand distortion, is well documented across many sectors and affects numerous companies, economically and environmentally, every day. Some of the main reasons for demand distortion are as follows:

- Time lags between the downstream customer pull and the upstream replenishment, including delays due to physical processing lead times, transportation, and waiting for information processing systems, such as waiting for the run of a material requirement planning system
- Multiple decision points where orders can be manipulated by individuals or system policies
- Other reasons such as minimum batch size production policy or disturbances to replenishment capability such as machine breakdown, and so on

The above systemic reasons lead supply chain operators to manipulate the orders in the wrong direction, to mistake a temporary peak in sales for a more prolonged change in demand, to panic and overestimate or sometimes to underestimate the customer requirements.

Furthermore, since there are several decision-making and ordering points along the whole supply chain, this distortion in the understanding of the actual demand is accentuated further up the chain, as several decision makers multiply the degree of over- and underestimation. The situation arises in the first place because customer demand will always vary naturally even if just a little, as it is influenced by factors that are to some extent predictable and others that are not. What we buy is influenced by a huge number of factors such as the weather, local events, sporting occasions, the showing of a popular TV program or even workers digging up the pavement.

Interestingly, however, despite all these potential variables, consumer behavior does exhibit significant patterns in relation to certain cycles. For example, in Figure 7.6, the *indexed variation*[45] in real customer demand for the ready-meal product is 26%, while the indexed variation of production of the sleeves at the packaging manufacturer is no less than 544%. The phenomenon has been visualized by adding representative lines relative to the size of the variation on the right-hand side of the graph.

Another contributory factor to the amplification and sometimes exponential distortion of demand is that while orders take a long time to be processed, retailers sometimes make unexpected alternations to their requirements, perhaps in response to an unanticipated buying pattern produced by an unexpected weather cycle. For this reason, suppliers in the FMCG sector tend to overproduce in order to have a buffer stock ready to be able to meet a sudden upward shift in demand. In some cases this strategy works very well. However, in other cases it can generate waste.

Demand distortion is one of the key underlying reasons driving product waste in supply chains. It drives big fluctuations in the supply chain, which in turn drives suppliers to create buffers of finished goods, raw material, or excess capacity. As explained previously, all of these buffers can lead to economic and environmental waste.

Therefore, the supply chain improvement team acted upon mechanisms for reduction of the demand distortion phenomenon such as sharing demand data, reducing time lags in the supply chain structure (i.e., reducing product lead time in manufacturing and distribution), more smooth ordering patterns, and a more rigorous forecasting system. The team also considered more radical solutions between the packaging supplier and the ready-meal

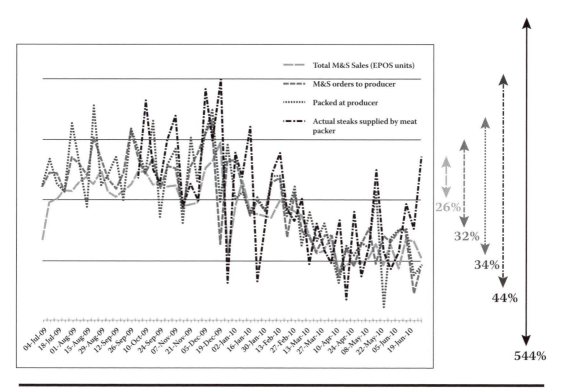

Figure 7.6 Demand amplification map.

manufacturer to both reduce the amount of packaging goods and eliminate the possibility of wastage due to frequent packaging design changes. The team is currently considering putting in place a vendor-managed inventory system with the packaging suppliers. All in all, the attempts to dampen the demand distortion effect had positive impact on both leanness and greenness of the supply chain.

Benefits of Lean and Green Improvement along the Whole Chain

Given the hands-on nature of the intervention, there were many quick wins identified during the project such as the realignment of product specification to the consumer needs and reduction of the giveaway meat waste. In total, around 65 opportunities were identified for improvement during various mapping sessions and workshops. Out of these opportunities, the team of senior stakeholders created a list of 16 vital projects based on potential cost–benefit analysis and allocated appropriate resources to implement each project. In order to operationalize these projects, a supply chain continuous improvement office was established to manage implementation of all projects while enabling internal staff to take ownership of the projects and sustain change throughout the chain.

One of the key tasks of the office is to roll out the findings to other manufacturers and suppliers in the coming years. It is also tasked to train staff wherever required. It is expected that the benefits, assuming all vital projects are implemented, will amount to several million pounds in cost savings while availability, consumer satisfaction, and environmental performance should increase significantly. No major investment has been required so far. At the same time, it is estimated that the environmental savings accrued from implementing only some of the vital projects, so far, is in excess of 880 tons of CO_2 or equal to driving more than 6 million kilometers in an average-sized automobile.

A key task of the supply chain continuous improvement office was to establish and monitor a set of agreed upon supply chain KPIs across the whole chain. This allows fact-based management and leads to more objective collaboration to create even more win–win solutions. Table 7.3 illustrates a set of KPIs suggested by the improvement team to be monitored. Here it can be seen that from the current state situation, the team outlined two stages of development: the first phase to stabilize the current state with more modest improvement targets over around 6 months and the second phase to deliver more significant improvement targets in up to 18 months.

Lean and Green Supply Chain Management: Case Study of MAS

World's First Carbon-Neutral Bra in Sri Lanka

As mentioned in the M&S case study, one of Plan A commitments was to establish model eco-factories in the M&S supply base. The first in Sri Lanka to be designed in line with the Plan A framework was the MAS Intimates Thurulie factory (Figure 7.7) run by MAS Holdings which, according to Vidhura Ralapanawe, the current manager of sustainability, has had a long-standing ambition to be at the forefront of people engagement and environmental performance.

MAS manufactures lingerie for international brands such as Victoria's Secret, Nike, and M&S. In collaboration with Marks and Spencer, it created the world's first-ever carbon-neutral bra. As simple as it may appear, a bra can be a remarkably complex product, into the making of which goes more

Table 7.3 Supply Chain KPIs and Target Improvement

KPI	Unit	Current State	Stabilize Target (6 Months)	Improve Target (18 Months)
On-shelf available	% (measured through mystery shopper method)	—	—	—
Shelf life	Days of life of product once replenished onto retail shelves	—	—	—
Waste	% (total value lost in the end-to-end chain)	—	—	—
Emissions reduction	Tons of $CO_{2\text{-eq}}$	—	—	—
Throughput time	Hours from manufacturer to shelf	—	—	—
Demand Amplification Index	Coefficient of variance of packaging manufacturer/ Coefficient of variance of shopper demand	—	—	—
Touches	Number of times product touched by human hands from manufacturer to shelf	43	35	14
People engagement	Qualitative measure (1–5)	—	—	—

than 20 different pieces and several packing materials, each one of which is typically sourced from a number of different suppliers. In order to create the carbon-neutral bra, MAS and M&S took into account the carbon footprint of all of the different components in their entire life cycle, from manufacturing to customer use, and set out to balance it by means of different innovative solutions. The first step was to radically reduce the carbon footprint at the manufacturing stage by designing a state-of-the-art model eco-factory in Thulhiriya, Sri Lanka.

Located in the MAS Fabric Park in Thulhiriya—which offers production, warehousing, residential, commercial, as well as cultural and social facilities—the civil construction of the factory cost $2.66 million, where M&S provided $400,000 to pay toward eco-design features. The building, which was

Figure 7.7   MAS Intimates Thurulie factory in Sri Lanka.

condensed onto two floors taking up only 6,780 square meters, incorporates significant sustainability technologies and has been highly acclaimed. In his report, Professor Leibundgut,[46] professor of building services at the Swiss Federal Institute of Technology, commented: "As an engineer I have seen uncounted industrial buildings in my lifetime, but no project has impressed me as favourably as this two-story building."[47]

As with other examples featured in this book, such as the Adnams distribution center, the construction of the factory incorporated eco-materials, in this case compressed stabilized earth blocks manufactured 40 kilometers from the site. Unlike Adnams, the walls did not require a plaster finish and are simply sealed with varnish on the interior and exterior. During construction, debris was channeled into the foundation of paving and most of the other waste was recycled in order to reduce the amount going into landfill. Energy sustainability has also been considered in how the factory will operate. The building incorporates passive cooling to reduce energy consumption in the hot Sri Lankan climate, achieved through the use of green and photovoltaic roofs as well as the primary cooling of 1–2°C achieved by increasing the density of surrounding vegetation. The green roof absorbs heat without transmitting it into the building, while the photovoltaic roof reflects most of the excess solar energy that is not transformed into electricity. This partly explains why the electrical energy required for running the factory is 25% lower than that of comparable factories. Furthermore, to supply the environmentally efficient cooling systems, an offsite hydro-electric facility supplies 90% of the required energy while the photovoltaic panels cover the rest.[48] For these reasons, it is believed that this facility is the world's first clothing factory powered solely by carbon-neutral sources.

The building is carefully situated on an intensively planted site to minimize its footprint and maximize open space, leaving a preexisting pond and the dense woods on the western part of the site undisturbed. The factory has a rainwater capture system, which reduces the consumption of potable water by half. All roads and pathways are paved with porous, cement-stabilized earth to reduce runoff and help recharge the groundwater, and the facility incorporates an anaerobic digestion system for sewage treatment, which also supplies methane gas for cooking in the kitchen. Many of these are suggestions of Marks and Spencer's Plan A that have also led to attainment of various accreditations for the building.[49] Furthermore, in line with Plan A objectives, the factory strives toward reducing waste sent to landfill to zero, and so provisions are made for the recycling of empty thread cones, paper, plastics, glass, and metal. Currently waste fabric is shipped to China for reuse and recycling purposes through waste vendors who buy the fabric locally.

The collaboration between MAS and M&S was not focused just on creating an environmentally friendly manufacturing site, but the emphasis was on implementing and operationalizing a truly sustainable site meeting the triple objectives of environmental preservation, people and societal performance, and economic longevity. This means addressing the more complex two-way or three-way relationships among the environmental, social, and economic performance of the site. For example, according to Vidhura, who was the team leader for the building project and contributed several essential concepts to the final plans, right from the very beginning, the design team "made a conscious choice to optimize the building–human ecosystem rather than just the building."[50] This means that the structure encourages behavioral changes. Requirements for human behavioral changes include always wearing climate-appropriate clothing and being a conscious consumer of energy, for instance, by switching off lights when they are not needed. This might seem strange given the objective of improving environmental performance. However, Vidhura explains, "We could have designed a motion-sensitive lighting system that automatically switches off when no one is around. But if one gets used to switching off an unnecessary light at work, he or she will do the same thing at home. And we know this is actually happening."[51]

Indeed, while all members of the workforce receive detailed training that explains the need and processes to be kept in mind, the factory has also hosted more than 1,000 visitors interested to learn from its operation. "We designed the factory to be a centre for learning and inspiration," Vidhura says. "[I]f you're serious about sustainability you have to share your ideas."[52] It is in this way that Vidhura is realizing his objective to "redefine the frame of what sustainability means."[53]

The Thurulie factory is also designed to directly promote worker welfare. Large windows allow views into the green surroundings from all the working spaces. The factory has showers, toilets, and lockers for employees, as well as a canteen (which is naturally ventilated and allows workers to eat free meals next to the pond), an infirmary, and on-site banking. As would be expected, the facility holds ISO 14001 and OHSAS 18001 certification for occupational health and safety management.

The synergies between the building architecture and human activities are also manifested in that the "worker-centered design" of the model eco-factory also allows "industrial processes [to] follow the high-productivity MAS lean manufacturing standard." To begin, the building consists of production floors separated into three halls on two floors. This makes climate control more efficient and allows the complete value stream to operate in each hall, consistent with lean manufacturing concepts. These production spaces do not have columns or other obstacles, so that each production team can arrange its machinery to best suit the garments being made, if necessary, in a production cell.

Electrical receptacles and lighting fixtures are also provided in a way that allows them to be used flexibly. Given the emphasis on production line flexibility and collaborative ways of working in lean manufacturing, there are spaces designed for impromptu meetings, which do not isolate the workers from one another. The company calls these quiet, open work areas *relax stations*. More broadly, the design of the building provides a visual connection between offices and production floors, consistent with the lean concept of Genchi Genbutsu (go and see) and implies a deep connection between office-based managers and the shop floor.

Another key feature of lean manufacturing in the MAS production system is just-in-time manufacturing of garments, which dramatically reduces the production lead times while also improving quality and reducing the need for storage space, leaving more area for production. The internal layout incorporates just-in-time practices while minimizing transportation of the goods around the building.

While the adaptable infrastructure has greatly facilitated implementation of lean in various aspects of operation at Thurulie, Vidhura reminds us that just looking internally is not sufficient. Instead, it is necessary to look at the supply chain as a whole. Starting in 2004, MAS began to apply lean thinking, first in their manufacturing processes and more recently in the front end of their business operations. During this process, a whole range of lean tools have been deployed with good effect.

The company has facilitated the spread of a culture of continuous improvement and a very robust Kaizen culture, including annual exhibitions showcasing the staff's continuous improvement achievements. Conversion of hierarchies into self-managed teams, in-line quality checks, and multiskilling of all operators have been key parts of creating a lean and green enterprise at MAS Thurulie. All teams now do their own weekly planning, and they deploy structured problem-solving methods such as fishbone diagrams. In sum, Vidhura says that now, "rather than having disempowered people doing what they can in their limited part of the process, the workforce actively and collectively takes ownership of planning and production. We have instilled the pride of workmanship in our team members."

The result of lean manufacturing can be seen, for example, in the reduction of the dock-to-dock times from 56 days down to 23 days. The major wins from MAS's lean journey include significant enhancement of employee engagement, reduction in lead times, reduction in inventory and warehouse space, optimal use of factory space leading to increased capacity in production from the same space, productivity improvements, and cost reduction. However, what MAS is most proud of is the way that issues are identified and then solved by the workforce itself.

Indeed, at the recent MAS Innovation Exhibition, 2000 square meters of floor space were covered with the outcomes of a collective culture of continuous improvement. At the same time that some of these concentrate on making processes more economically efficient, others focus on green issues. For example, faced with the requirement by brand owners to burn unused branded fabrics, one team implemented a solution in which they use a fabric shredder, which now allows the waste to be sent to recycling.

As with the carbon-neutral beer developed by Adnams, it was still necessary to offset the remaining emissions after all possible reductions along the supply chain. To achieve this, the team partnered with Rainforest Rescue International to build a biodiversity corridor between the Polgahakanda and Kinneliya Forest Reserves. In total 6,500 trees, all indigenous species, were planted to physically interlink the two existing canopies. The trees chosen,

Figure 7.8   Visual management and visual control. Online production status and discussion on data at floor.

however, have little or no value as timber, which provides a disincentive for local communities to cut them down. Also, 40% of the trees are income bearing by producing spices and fruits. These two factors, combined with the stipend that is paid to farmers for their safe maintenance, ensures that the initiative has sound economic foundations. However, despite the success of this program, Vidhura and his team are presently exploring the possibilities of further reducing the embodied carbon, for example, by substituting the current materials for regenerated cellulose and recycled materials.

All in all, the model eco-factory project, by incorporating various lean and green features, has been an economic success for MAS, while also enhancing the images of both MAS Holdings and Marks and Spencer. Indeed, Vidhura is candid to say, "If we didn't do lean we wouldn't be in business or we would be in a pretty bad shape by now." This is because the costs of manufacturing in Sri Lanka are higher than those in many other Asian countries, and where most apparel companies have found it difficult to survive, the lean and green approach has given MAS the efficiency and the productivity needed to thrive. Although M&S provided

Figure 7.9   Waste fabric collector with combined band-knife cutting machine—5S at Sustain stage.

the impetus for the construction of the factory, the 25% price premium required for the sustainable technology was only partially met by M&S's financial contribution of $400,000. This meant that MAS had to invest significantly while receiving no premium on the product bought by the UK retailer or any guarantees to purchase the merchandise produced. For this reason, making a success of the factory has been down to MAS Thurulie, which has specialized in sophisticated, high-value lingerie garments and commands a unique selling point.

In conclusion, the achievements in this story, in the light of the impetus provided by Plan A, demonstrate how the triple objectives of sustainability can be delivered by means of collaboration and fact-based management in the supply chain. Indeed, the story of MAS Thurulie has inspired many M&S suppliers to incorporate innovative solutions for sustainability into their own products and processes. At the same time that this specific project resulted

Figure 7.10 Cut to box production: Sewing lines are catered by cutting table on the floor; reduced transportation and reduced WIP with visual controls.

in the carbon-neutral bra, the knowledge generated has also been used in other new products stocked by M&S.

Moreover, Vidhura Ralapanawe has gone on to be voted as Honda's 11th Cultural Engineer, a huge accolade and a fantastic testamony to his lifelong personal commitment to providing incrementally more impactful solutions to some of the world's biggest problems.

Interestingly, when we asked Vidhura if lean is green, he sharply replied "*No.*" He then went on to explain that it all depended on the way lean and green are being implemented, for example, where hot molding machines were moved in-line to reduce lean wastes (e.g., unnecessary movement) at an MAS factory, this meant that their head had to be extracted individually, which increased the environmental impact. Here, Vidhura says, "managers have to make lean be green."

Managers must reframe their aims and combine green and lean together for simultaneous economic and environmental results—something that he is still working on tirelessly to achieve in the Thurulie factory and beyond.

Conclusion: Role of Lean in Delivering Plan A

This chapter, so far, has reviewed the application of lean and green approaches along the extended supply chain by one of the world's most sustainable retailers, Marks and Spencer. Starting with a review of the company's history, it emerged that the principles of quality, value, service, innovation, and trust have long been central to operating practices. Slowly, M&S has increased the scope of stakeholders to whom they have felt a level of responsibility: from a concentration on the welfare of their workers, through to an increased emphasis on community, and finally to a global scale in the development of their landmark Plan A initiative.

This journey has also included a growing realization that the issues of ethical work practices, environment, and profitability are intimately intertwined. For example, in the case of general merchandise production, the experience of M&S and their supply partner, MAS Intimates, has been that through facilitating greater employee engagement and empowerment, creative solutions emerge for problems that may never have been noticed. We will expand on this relationship from the MAS point of view in the following text to show how lean and green can be adopted, equally successfully, in international supply chains and in the developing world.

Furthermore, M&S has begun to develop ways to lead lean and green thinking into its wider supply base. The lean and green intervention across the extended supply chain that was explained in this chapter is an excellent demonstration of how lean (economic) improvements drive environmental benefits and vice versa.

Since they are partly a brand house and partly a retailer, it is critical to M&S to have the full engagement of their suppliers. In fact, Plan A targets are only achievable if the suppliers are fully aware of, on board with, and engaged with the program. In the case of food supply, M&S is deploying the Food Supplier Sustainability Framework scorecard, which integrates environmental, ethical, and lean thinking aspects. The impact, so far, has been greatly improved environmental and ethical performance as well as cost savings and increased productivity, which have brought better returns to all stakeholders along the supply chain.

Tracing the background from which the application of lean methods first emerged within M&S, this case study has focused on how these analytical tools have contributed to the Plan A program. Specifically, this chapter has provided a detailed account of how the application of lean thinking to a selected SKU allowed the company to transcend the common limitation of looking only within the organizational boundaries to focus instead on lean and green across the extended supply chain with remarkable success.

One of the strongest narratives from the experiences of M&S is that lean has been a great enabler for realizing Plan A objectives, while at the same time, Plan A has been the best facilitator for the attainment of the fundamental and cultural changes required to create a lean enterprise and to extend it to the wider supply chain. Today, lean is an integral part of Plan A and they are equally regarded as key sources of bottom-up innovation and continuous improvement.

Endnotes

1. Information used in this section has been obtained from publicly available sources including the M&S website and sustainability reports. Further information has been gathered from interviews with M&S staff and firsthand research, which are clearly marked throughout the case study. However, this case study reflects the authors' own opinions.
2. Marks and Spencer. 2012. Plan A: Doing the right thing. Your M&S. http://plana.marksandspencer.com/about
3. Interview with Carmel McQuaid, February 2012.
4. Ibid.
5. Wills, J. 2011. M&S—Wholly embracing staff in plan to become the world's most sustainable retailer, The Guardian, May 26. http://www.guardian.co.uk/sustainable-business/staff-plan-worlds-sustainable-retailer.
6. Ibid.
7. Ibid.
8. Nichols, W. 2010. Mike Barry: M&S is cashing in with Plan A. BusinessGreen, November 9. http://www.businessgreen.com/bg/interview/1868061/mike-barry-cashing-plan
9. Bense, C. 2011. Meet the candidates. Food Manufacture, June 1. http://www.foodmanufacture.co.uk/People/Meet-the-candidates

10. Marks and Spencer. 2011. How we do business report. http://cor-porate.marksandspencer.com/documents/publications/2011/how_we-do_business_report_2011

11. Felsted, A. 2011. Marks & Spencer's green blueprint. Financial Times, March 17. http://www.ft.com/intl/cms/s/0/0a77b918-4404-11e0-8f20-00144feab49a.html#axzz2ED4Yv9DO

12. Marks and Spencer. 2011. How we do business report. http://cor-porate.marksandspencer.com/documents/publications/2011/how_we-do_business_report_2011

13. Brand Sustainable Futures, 2010, http://www.havasmedia.com/2010/10/retailers-poised-to-become-the-new-googles-of-the-consumer-goods-industry/

14. Marks and Spencer. 2011. How we do business report. http://cor-porate.marksandspencer.com/documents/publications/2011/how_we-do_business_report_2011

15. Ibid.

16. Ibid.

17. Ibid.

18. Ibid.

19. Leave Your Hanger with us. Your M&S, 2012. http://plana.marksandspencer.com/about/partnerships/unicef.

20. Marks and Spencer. 2011. How we do business report. http://cor-porate.marksandspencer.com/documents/publications/2011/how_we-do_business_report_2011

21. Leibundgut, Hansjürg. 2009. Clothing factory in Sri Lanka. Zurich: Holcim Foundation for Sustainable Construction. http://www.holcimfoundation.org/Portals/1/docs/Book_MAS_SriLanka.pdf

22. Smriyi, D. 2011. In conversation with . . . Vidhura Ralapanawe: Going green. Sunday Times, Sri Lanka, October 23. http://smritidaniel.wordpress.com/2011/11/20/vidhura-ralapanawe-going-green/

23. Interview with Carmel McQuaid, February 2012.

24. Marks and Spencer. 2011. How we do business report. http://cor-porate.marksandspencer.com/documents/publications/2011/how_we-do_business_report_2011

25. World Wildlife Federation Seafood Charter. http://assets.wwf.org.uk/down-loads/seafood_charter.pdf

26. Marks and Spencer. 2011. How we do business report. http://cor-porate.marksandspencer.com/documents/publications/2011/how_we-do_business_report_2011

27. Interview with Carmel McQuaid, February 2012.

28. Interview with Louise Nicholls, February 2012.

29. Anonymous factory manager quoted in interview with M&S staff.

30. Interview with Louise Nicholls, February 2012.

31. Ibid.

32. Interview with Carmel McQuaid, February 2012.

33. Leibundgut, Hansjürg. 2009. Clothing factory in Sri Lanka. Zurich: Holcim Foundation for Sustainable Construction. http://www.holcimfoundation.org/Portals/1/docs/Book_MAS_SriLanka.pdf
34. Food Chain Centre. 2007. FCC Completion Report. United Kingdom: IGD.
35. Ibid.
36. Through involvement with two of the authors, both project leaders of the farm-to-fork studies at Cardiff University.
37. Normative learning includes changes in values or behavioral norms as a result of empirical observations of the surrounding environment.
38. Zokaei. 2008. A systemic analysis of the UK food supply chains. PhD diss., Cardiff University.
39. In this case, the index for demand amplification is calculated as 2 × standard deviation over mean, and is meaningful only as a relative or comparative measure of distortion of real consumer needs.
40. Leibundgut, Hansjürg. 2009. Clothing factory in Sri Lanka. Zurich: Holcim Foundation for Sustainable Construction. http://www.holcimfoundation.org/Portals/1/docs/Book_MAS_SriLanka.pdf
41. Ibid.
42. Ibid.
43. The building is compliant with the U.S. Green Building Council (USGBC) standards for green buildings and complies with Leadership in Energy and Environmental Design (LEED) Platinum standards.
44. Ibid. p. 68.
45. Ibid. p. 69.
46. Smriyi, D. 2011. In conversation with Vidhura Ralapanawe: Going green. Sunday Times, Sri Lanka, October 23. http://smritidaniel.wordpress.com/2011/11/20/vidhura-ralapanawe-going-green\
47. Interview with Vidhura Ralapanawe, February 27, 2012.
48. Leibundgut, Hansjürg. 2009. Clothing factory in Sri Lanka. Zurich: Holcim Foundation for Sustainable Construction. http://www.holcimfoundation.org/Portals/1/docs/Book_MAS_SriLanka.pdf
49. Interview with Vidhura Ralapanawe, February 27, 2012.
50. Ibid.
51. Ibid.
52. Leibundgut, Hansjürg. 2009. Clothing factory in Sri Lanka. Zurich: Holcim Foundation for Sustainable Construction. http://www.holcimfoundation.org/Portals/1/docs/Book_MAS_SriLanka.pdf
53. Interview with Vidhura Ralapanawe, February 27, 2012.

THE WAY FORWARD

Chapter 8

Lean and Green Growth: An East Anglian Model

The world has been making commitments to facilitate a lower-carbon economy for decades. From the Report of the Brundtland Commission,[1] *Our Common Future*, published in 1987, through to Rio+20,[2] we have long been grappling with economic growth, environmental protection, and social equality.

But it has not happened yet, and it is hard to predict when it will, or what a new, greener future will look like, without making it happen our-selves. In the words of Peter I. Drucker, "the only way to predict the future is to invent it."[3] Local, bottom-up initiatives that are led by communities are where lasting change will occur and that is our challenge.

In the United Kingdom, the counties of Norfolk and Suffolk in East Anglia are weathering the financial storm quite well. They possess some of the richest farmland in the European Union (EU), a number of high-tech clusters, a healthy diversity of energy sources, plus a wealth of cultural sites and a strong tourism industry. But they also face challenges. As the driest, lowest-lying region of the United Kingdom with an extensive coastline and a large agricultural sector, East Anglia is literally on the front line of climate change. So any growth in the economy has to be green growth.

The area has been chosen by the UK government (Green Economy Pathfinder) to lead on a more sustainable model for growth, both in green sectors such as low-carbon energy generation and climate change adaptation services, and others, as they open up to green goods and services. The area is bringing together academics, businesses, and the public sector under a recently established Local Enterprise Partnership to support and develop the following:

▪ Energy and resource efficiency to bring to market opportunities, cost savings, and increased competitiveness
▪ Innovation to sustainably lower costs and improve performance
▪ Investing in infrastructure so that transport, energy, and communication networks support the move to a green economy
▪ Resilience in business planning to adapt to the challenges and risks of energy and resource security, fluctuations in fossil fuel prices, and the impacts of climate change
▪ Work with marketers and consumers to enable them to make informed greener choices through relevant information and clear communication

What Is the Lean and Green Economy?

A lean and green economy is one in which economic growth is combined with continued reductions of greenhouse gas emissions and other environmental impacts. Through effective demand management and efficiency measures, energy, water, and other natural resources will be used efficiently, inputs to the production processes are optimized, and the level of waste to landfill decreases or is eliminated completely. The UK government's transition to a green economy paper *Enabling the Transition to a Green Economy: Government and Business Working Together* proposes the following definition:

> A green economy is not a sub-set of the economy at large—our whole economy needs to be green. A green economy will maximise value and growth across the whole economy, while managing natural assets sustainably. It will help UK businesses take advantage of new markets for environmental goods and services, and to demonstrate the strong stance the UK is taking internationally to reduce carbon and tackle climate change.

By reducing reliance on fossil fuels and transitioning to more renewable energy, any economy will be more resilient. A prosperous and thriving lean and green economy could generate the investment, innovation, and skills needed to transform products and services, develop cleaner technologies, and help capture new markets for businesses and regions within the UK and internationally. Countries, regions, and businesses that are attuned to these trends will be well placed to exploit these comparative advantages and benefits as the market for greener goods and services expands. However, to create a truly sustainable economy, growth needs to be linked to positive impacts upon the environment with resources being used not only efficiently but also intelligently. Efficiencies will not necessarily help stem our negative impacts if our need for resources continues to rise. A functioning green economy will both increase efficiencies and stimulate the development of technologies and systems that would enable a truly sustainable economy.

In a lean and green economy, companies continuously innovate to eliminate non-value-adding activities across their internal operations as well as along the whole supply chain working with their suppliers and customers. This leads to better resource efficiency as well as enhanced consumer satisfaction; that is though if the requirements of consumers are not in conflict with the notion of sustainability. Government mechanisms are essential to reinforce sustainable consumption. In a lean economy, at least in theory, companies will deliver whatever is considered consumer demand. However, in a lean and green economy, customer value is aligned with the goals of sustainable development.

From society's point of view, organizations are there to create value and to solve consumers' problems. The real definition of lean is value enhancement for consumers and for the whole society, and not cost reduction. People are looking in the wrong end of the telescope when they think that lean is about cost reduction. The purpose of the organization, in a truly lean economy, is to maximize value to consumers and to the society at large.

Such an economy would have the following characteristics: It would see public procurement policies promoting the greening of business and markets through direct government purchases of more sustainable and efficient products. It would also see public investment in sustainable infrastructure, such as public transport, renewable energy, and retrofitting of existing infrastructure and buildings for improved energy efficiency. Fiscal structures would be aligned with environmental goals with the aim of getting prices "right" including removing subsidies, valuing natural resources, and imposing taxes on things that harm the environment. Strategic investment through the public sector would be the foundation of socially and environmentally sustainable growth.

Regional Context

With 60,000 enterprises and a population of 1.5 million, the counties of Norfolk and Suffolk are not without economic significance. Tourism and energy are two of the areas' major sectors bringing wealth and the potential for growth to the economy. The area has huge potential in terms of farm-land, landscape, and tourism, but faces pressure on water resources, food production, and housing. This said, the area is well-positioned to capital-ize on and prove the benefits of a low-carbon economy. Achievements to date have been both encouraging and impressive in the areas of low-carbon impact and sustainable growth and employment.

Low-Carbon Impact

After Aberdeenshire in Scotland, the east of England is the second largest center for the UK energy industry. Recent gas discoveries in the North Sea suggest there could be continued activity in the region and there are eco-nomic opportunities around adapting existing offshore assets for carbon cap-ture and storage. EDF Energy's Sizewell B provides 3% of the UK's electricity and has saved an estimated 67 million tons of CO_2 since 1995. And if, after public consultation, the planning for Sizewell C is successful, it will generate 3.2 GW of low-carbon electricity—enough for 5 million homes. Norfolk and Suffolk are also very close to planned developments for 20,000 MW of off-shore wind energy. The East Anglian array will be the largest development of its kind in the world.

The area also benefits greatly from the University of East Anglia's Low Carbon Innovation Centre and is stimulating the uptake of carbon-reducing technologies and practices. In Suffolk, some 300 Small Medium Enterprises (SMEs) have received energy efficiency audits identifying carbon savings of 7,200 tons and £1.2 million in energy savings. Energy Innovations, another Suffolk company, has supplied an off-grid district heating and hot water system that has significantly reduced CO_2 emissions at Snape Maltings, and a partnership with the Bio Group will create around 1.2 million cubic meters of biomethane as well as a natural liquid fertilizer and compost from brewery waste at Adnams PLC. This project will produce 9.6 million kilowatt-hours of energy and avoid 120,000 tons of CO_2 emissions every year. Low-carbon housing is also an area of focus and Greenright Homes' development in Wortham, Suffolk, is projected to reduce annual central heating running costs by thousands of pounds and more than halve carbon emissions compared to homes heated more traditionally by gas or electricity.

> A number of building and restoration projects have recently been completed in Norfolk and Suffolk that clearly demonstrate it is possible to deliver sustainability and incorporate low-carbon and renewable heating and energy solutions while still being commercially viable. Simon Bennett[4] of Greenbright Homes commented on creating environmentally friendly homes.

As for commercial buildings, the two projects at the University Campus Suffolk completed by Willmott Dixon both achieved BREEAM Excellent certificates through a combination of energy- and water-efficient design and construction solutions. The buildings incorporate automated building energy management systems, infrared sensor lighting, super low-emission boilers, "green roofs," rainwater harvesting, and water leak detection systems. The approach to this type of building project is best summed up by Martin Ballard,[5] Group Environmental Manager at Willmott Dixon Group:

> We play an active role across the built environment and aim to leave a positive legacy for communities and wildlife and create a greater sense of place for people living in and around the assets we create and maintain for our clients.

A real sense of place and a concern for the natural, built, and social environment is quite evidently high on the agenda of the whole area. Adapting old buildings and retrofitting renewable energy sources at a Grade 2 listed site such as Snape Malting is an excellent example of what can be achieved, even in a very aged building.

Local communities are also benefitting from a strong cooperative and social enterprise movement within the region. The East of England Cooperative is the third largest independent retailer in the country and many communities are self-organizing around the national transition town model, creating initiatives that build resilience to the pressures of climate change and fossil fuel reliance. Transition Ipswich plans to build a community-owned wind turbine at Thorington Barn that will generate 16.5 GWh of renewable electricity every year, enough to power 3,600 homes.

In terms of transport, a courtesy bus service from Ipswich Station to British Telecom's (BT) global innovation and development center at Adastral

Park gives business travelers a viable alternative to taxis for the seven-mile journey, saving money, fuel, and CO_2 emissions. Under the Suffolk Climate Change Partnership, many other such initiatives exist and are seeking to improve the area by installing facilities that enhance and encourage walking, cycling, and the use of public transport in an attempt to make Suffolk the greenest county in the UK.

Sustainable Growth and Employment

The Incrops Enterprise Hub is a virtual network of research, public sector, and corporate partners that collaborate on commercial uses for innovative crops and the research into plant science and microbiology at the John Innes Centre. Located on the Norwich Research Park, it benefits agriculture, the environment, and human health, and already contributes around £170 million pa to the UK economy. The nuclear energy sector has provided continuous employment in East Anglia since 1960, and the current operator EDF Energy employs more the 750 staff and contractors, including 50 apprentices. Although many are small, rurally located, off-grid energy companies are also beginning to make their presence felt in the local economy. The Bio Group employs 40 people and renewables specialist RenEnergy has more than 50 full-time staff. Not to be forgotten, tourism accounts for some 17% of businesses on the Norfolk Coast Area of Outstanding Natural Beauty (ANOB) and 16% of the Suffolk Coast and Heath ANOB. Combined, these natural businesses account for 3,400 jobs and highlight the ability of nature, tourism, and industry to work positively side by side.

To support this growth model, local universities and colleges are reintroducing degree courses in engineering and advanced manufacturing linked to the energy sector through the East of England Energy Group (EEGR) and their skills for energy initiative. This focus upon energy has also been extended into schools where initially a small number of pupils and teachers in schools across the area have become energy ambassadors and trained to lead energy reduction in their schools.

The combination of green initiatives being taken across East Anglia is attracting the attention of the investment community with the University of East Anglia's Low Carbon Innovation fund leading the way, and the area is seeing interest from larger London-based funds.

There are many regions and counties within the UK and Europe undertaking terrific work in the greening of the UK's economy, but initiatives are often isolated and lack a joined-up feel. Norfolk and Suffolk see establishing a joined-up, vibrant, and progressive green economy as a unique selling point and significant sustainable growth opportunity. Initially, many of the initiatives and projects are small in global terms; however, the combined counties do sense a once-in-a-lifetime opportunity. They recognize that bottom-up, local change is where sustained benefits will accrue and that this can be scaled up for the benefit of the wider UK economy. This Pathfinder project could be shown to deliver not only significant greenhouse gas reductions but also significant lean and economic benefits as the UK transitions to the cleaner, greener, and leaner future it needs.

Endnotes

1. Brundtland Commission, 1987. *Our Common Future*. http://www.un-documents.net/our-common-future.pdf
2. Rio +20, 2012. United Nations Conference on Sustainable Development. http://www.uncsd2012.org/content/documents/colombiasdgs.pdf
3. Drucker, P. F., 2001. *Essential Drucker: Management, the Individual and Society*. HarperBusiness: New York.
4. Bennett, S. Greenbright Homes.
5. Ballard, Martin. Willmott Dixon Group.

Chapter 9

Conclusions and Wrap Up

As you have read, bringing Lean and green together is an interesting challenge and one that a number of leading organizations that we have featured here have attempted to do. These companies have found their journey to be extremely rewarding. However, as it stands, the vast majority of organizations are still in their infancy on the lean and green endeavor. We hope that this book will help more to follow this path.

A common feature of those that have succeeded is that they have sought synergy and even better, symbiosis, between lean and green. They have sought to build their organizations (and sometimes their wider supply chains) as places living and breathing the mantra of *tomorrow better than today*. In addition, a common theme has emerged that there needs to be a much wider strategic focus than just the *voice of the owners* as described in much of the classic strategy literature. There also needs to be a stronger focus on

- the voice of the customer (the end consumer of the product or service),
- the voice of the employee (to ensure that we show respect to these individuals and help them to make their life better), and
- the voice of society (in terms of the their wider social and environmental needs).

Companies that are driven by a vision, such as Toyota, Tesco, or Adnams, are far more likely to reap the benefits of lean and green than companies that are driven purely by economic goals. We have also seen that there is no easy short cut or *silver bullet* for creating a lean and green organization and that applying a range of lean and green tools in an unsystematic way is not likely to produce good long-term results.

Presently, many organizations seem to be void of systematic methods for continuous green improvement. Many environmental projects are implemented as top-down initiatives, by and large skewed in the direction of one-off technical fixes or end-of-pipe solutions involving a limited number of people. Scattered green improvement projects are unlikely to leave a lasting cultural change. Having discussed the shortage of systematic improvement approaches for integrating green and sustainability into business processes, we looked at the Lean and Green Business System model as a structured and holistic means of greening businesses.

Today, the lean community is increasingly aware of the importance of deploying systematic improvement approaches, paying attention to the cultural and leadership issues and orderly technical methods such as value stream mapping or road mapping techniques. Companies such as Adnams, Toyota, and Marks and Spencer have been successfully drawing upon the experience of the lean community to design their ever more effective sustainability journeys.

However, there are a number of significant organizational challenges, many of which are peculiar to each individual organization. So what is the way forward for you and your organization? In order to help you set this direction, we suggest you think through the answer to the following initial questions:

- What are we currently doing with lean and green in our organization?
- Are they working together or against each other?
- What is working well currently?
- What is not working well currently?
- What are the gaps between the current state and what we would like it to be?

When you have thought through these initial questions, you might like to then start planning what you will do; here are some suggestions in terms of further questions to consider.

Are you ready for lean and green? If you are a lone voice in the organization in either the lean or green area, you may well be facing an uphill struggle to gain traction. It is important to seek out allies in the journey, perhaps people with a passion or those with an active interest. It is also useful to judge what the likely opposition (if there is any) is going to be and how it might be overcome.

Do you know where you are now and what you are trying to achieve? We introduced the stages of the lean and green maturity model at the end of Chapter 3, as a way of judging where you are on the journey. We suggest you think about this as well as how this fits with the wider aims of your organization in terms of its targets and goals such as cost saving, profitability, and growth. It will also be useful to identify the hot green topics in your organization in terms of the environment, for example, water, energy use, landfill, and use of hazardous or toxic materials.

Have you considered connecting the lean team's objectives with the green team's and vice versa? We discussed how lean and green are essentially one and the same thing. We also looked at how leading-edge companies have created a symbiosis of the two concepts. One of the key areas that can release sudden energy is to connect the lean and green teams within your organization. It is especially important for environmental managers to tap into the "free energy" offered by the continuous improvement offices to get green team members to benefit from the transformation experience and the toolkit available to your lean team.

On the other hand, we should engage our lean and continuous improvement team members so that they constantly see the green opportunities and the green measures (or "currencies" as discussed in Chapter 3). We need to enable them to quantify the impact of their transformational programs in green terms—well, at least to the basic level. We have seen numerous lean initiatives yielding rich green benefits; nonetheless, the benefits often remain ignored, not quantified, or simply not realized due to lack of awareness. Green benefits are abundant when you do lean projects. However, there needs to be conscious effort to realize the green benefits alongside the environmental benefits.

How are you going to get wider acceptance across the organization? This may well be the biggest and most important step in your journey, as the lone voice rarely succeeds. What is essential is that you can create a burning platform for change that the various key players can get involved in and be excited about. Over many years of experience, we have found that it is extremely constructive to do this with the senior team in the organization by way of a focused event, such as a lean and green leadership event, perhaps facilitated by an external expert.

What is your strategy for implementation? Once you have the senior team on board, it is necessary to develop a strategy for implementation. Here you have choices around a make-or-buy decision. Do you have the right skills in the organization or do you need to buy in support for your journey?

What are your lean and green targets over the next few years?
Do you have lean and green targets for this year? How about reducing your
energy bill by 20%? We looked at a basic lean and green toolkit for improvement in Chapter 4, where reducing utility consumption by 10–30% was often
achievable through simple interventions. Do you know how much money
and how many tons of CO_2 you will save?

Where should you start? There is no right or wrong place to start.
Sometimes it is best to undertake some limited pilot work to learn about an
integrated lean and green approach such as the Kaizen events we discussed
in Chapter 4. In other cases you might need to start with process diagnosis
or skills development. We suggest these decisions be made with the senior
management team.

What resources are you going to need? One of the most frequent
problems that we have encountered with organizations implementing a
lean and/or green approach is an underinvestment in the number of people
needed to implement the changes. Indeed, what we see in the best organizations is that such change ultimately becomes part of the day job for
everyone.

What is going to be in your lean and green roadmap? In order to
achieve success, we firmly believe that you should create a roadmap for
your journey that covers at least the next eighteen months involving the
major activities, roles, and responsibilities and a balance between the four
elements of the lean and Green Business System model, namely Process
Management and Toolkit, Leadership and People Engagement, Strategy
Deployment, and Supply Chain Management. A lean and green maturity
assessment is sometimes the first step toward developing your lean and
green roadmap.

How are you going to undertake the improvement activities? Here
it is necessary to develop a more fine-tuned plan for each activity step, as well
as how you are going to deploy your internal or bought-in resource team.

How do you know whether it is working? During your implementation, it will be key to manage the change and see how well you are doing.
We find that the best way to do this is via some form of visual management
rather than a computer-based program, which often feels owned only by the
person with the keyboard.

How to make the approach stick? This is probably the most difficult
step and one that is well documented in the Staying lean text.[1] As a result,
we offer in Table 9.1 a set of tips; you might also like to think about how

Table 9.1 Ten Tips for Lean and Green Stickability

Tips for Lean and Green Stickability	How Are You Doing?
Think of lean and green as a philosophy for success rather than a series of tools and techniques.	
Apply lean and green across the organization, not just in the parts the textbooks talk about.	
Focus on improving whole processes not just individual departments.	
Link everything you do to create value for your customers, your organization, your people, and wider society.	
Don't just copy others; think through your approach, based on what you are trying to achieve.	
Make everyone aware of what you are trying to achieve and why.	
Align your communication and key performance measures to create and sustain a lean and green organization.	
Provide sufficient resources in terms of people and training across the organization, not just your improvement agents.	
Leaders need to not just talk about lean and green; they need to demonstrate in their actions that they are serious.	
Make sure your finance and rewards and recognition system appropriately encourage lean and green activity and motivate your people.	

Source: Adapted from Hines, P., P. Found, G. Griffiths, and R. Harrison R. *Staying Lean: Thriving, Not Just Surviving*. Cardiff, UK: Lean Enterprise Research Centre, 2008.

well you are doing perhaps every six months along your lean and green journey.

We hope you have enjoyed reading the book and we would all wish you good luck on your lean and green journey.

Endnote

1. Hines, P., P. Found, G. Griffiths, and R. Harrison. 2008. *Staying Lean: Thriving, Not Just Surviving*. Cardiff: Lean Enterprise Research Centre. http://www.qualitycoach.net/products/staying-lean-thriving-not-just-surviving-free.asp

Appendix A

Here is a more detailed overview of some of the key strategies portrayed by industrial visionaries for generating a new sustainable industrial era. Together they provide a vision of a sustainable future for the industry. Without one such vision and without a clear understanding of their favourable conditions, it is not possible successfully to implement the provisos of sustainability. What is common among all these strategies is the acceptance of the economic and developmental needs of society and a free, yet ethical. While the adoption of sustainability is essentially in its infancy, these visionary strategies provide important practical solutions. Clearly there is still a gap in terms of implementation and application in practice. Then again, in this book we draw upon a number of best practice examples to illustrate their applicability. The following table helps to explain what sustainable industry is or how it might look.

Table A.1 Key Strategies for Creating More Sustainable Industrial Practices

Strategies (School of Thought)	Key Contributions	Description
Industrial Ecology	*The Arrogance of Humanism* (Ehrenfeld, 1981) *Industrial Ecology* (Graedel and Allenby, 2003)	Industrial ecology (IE) is a framework for thinking about and organizing human economic and social systems in ways that resemble the natural ecosystems that emerged in the 1980s. IE is the study of the flows of materials and energy in industrial and consumer activities to investigate the effects of these flows on the environment, and the influence of economic, political, regulatory, and social factors on the flow, use, and transformation of resources. The uniqueness of IE is that it aims to show how environmental concerns can be integrated into economic activities. At the application level, IE offers tools for analysis of the interface between industry and environment, and provides a basis for management of environmental impacts.
Eco-efficiency	*Changing Course* (Schmidheiny, 1992)	The concept of eco-efficiency was introduced by the World Business Council for Sustainable Development (WBCSD) in the 1990s. Since then, eco-efficiency has gained increasing attention across businesses as well as among policy makers. The concept is based on creating more while using fewer resources and creating less waste and pollution. The seven critical factors for eco-efficiency are reduction of the material intensity of goods and services, reduction of the energy intensity, reduction of toxic dispersion, enhancing material recyclability, maximizing sustainable use of resources, reduction of material durability in nature, and increasing the service intensity of products.
Eco-effectiveness	*Cradle to Cradle* (McDonough and Braungart, 2002)	Eco-effectiveness is based on a cradle-to-cradle or closed-loop design strategy rooted in the systems of the natural world, which are not necessarily efficient, but are effective since there is no waste in the whole natural system, whereas each individual subsystem is creating waste. The principles of eco-effectiveness are waste equals food (create a closed-loop industrial system), use the current solar income, and celebrate diversity. Eco-effectiveness calls for transformation of human industry through ecologically intelligent design. It seeks to design industrial systems that emulate nature where waste from one subsystem is food for another. So, the tenet of eco-effectiveness is that waste equals food. Eco-effectiveness argues that eco-efficiency works with the same system that caused the problem in the first place, merely slowing it down with moral proscriptions and punitive measures.

| Natural Capitalism | *Natural Capitalism* (Hawken et al., 1999)
Factor Four (von Weizsacker, 1998)
Ecology of Commerce (Hawken, 1994)
The Value of the Ecosystem Services (Costanza et al., 1997) | Natural capitalism pictures a new industrial system based on a very different mindset and values than conventional capitalism. Natural capital refers to the natural resources and ecosystem services that make economic activities possible. *"Capitalism as practiced is a financially profitable, non-sustainable aberration in human development. What might be called "industrial capitalism" does not fully conform to its own accounting principles. It liquidates its capital and calls it income. It neglects to assign any value to the largest stocks of capital it employs—the natural resources and living systems as well as the social and cultural systems that are the basis of human capital".* (Hawken et al., 1999, 5). Natural capitalism is based on four strategies as offered by Hawken, Lovins, and Lovins in the late 1990s

1. Radical Resource Productivity: Obtaining the same amount of utility from a product or process while using less material and energy in order to slow resource depletion and pollution at the same time. Radically increased resource efficiency—at least in theory—lowers costs for business and society.

2. Bio-mimicry: Reducing the waste from product life cycles can be accomplished by redesigning products and processes as biological analogues. There is no waste in nature; waste from one process is food to another. Bio-mimicry means imitating natural processes and products. This changes the nature of industrial processes and products, enabling the constant reuse of materials in continuous closed cycles, more use of compostable products, and eliminating toxicity.

3. Service and flow economy: This strategy calls for a fundamental change in the producer–consumer relationship wherein services replace physical goods where possible. In this situation, the service provider is responsible for the product after use, and it is to the benefit of suppliers to make products last longer. So, less waste is generated in this type of relationship. Also, more jobs are created since the service industry is more labor intensive. In a service economy, customers have more choice since changing service providers is easier and cheaper than obtaining new goods.

4. Investing in natural Capital: Reversing worldwide planetary destruction by reinvestments in sustaining, restoring, and expanding stocks of natural capital, so that the ecosphere can produce more abundant ecosystem services and natural resources. |

(continued)

Table A.1 Key Strategies for Creating More Sustainable Industrial Practices (continued)

Strategies (School of Thought)	Key Contributions	Description
The Natural Step	*The Natural Step Story* (Robert, 2002) *Tools and Concepts for Sustainable Development* (Robert, 2000) *Dancing with the Tiger: Learning Sustainability Step by Natural Step* (Nattrass and Altomare, 2002)	The work on The Natural Step (TNS) framework was initiated by Professor Robert in the 1980s and received wide attention from industry and policy makers during the late 1990s and 2000s. TNS deploys a systems approach to describe the whole biosphere–society system in a way relevant to businesses and governmental decision makers. The systems approach helps to avoid misunderstandings and disallows intellectual escape routes. TNS looks at sustainable development (SD) at three levels: principles of ecosphere governed by natural laws of physics, principles of sustainability (four system conditions), and the principles for a process to meet principles for sustainability (the transition toward sustainability and then the safe development thereafter). The (TNS) framework's definition of sustainability includes four system conditions (scientific principles) that must be met in order to create a sustainable world. The first three principles are defined as the following: in a sustainable society, nature is not subject to systematically increasing concentrations of substances extracted from the earth's crust, concentrations of substances produced by society, and degradation by physical means. The fourth principle argues that in a sustainable society, human needs are met worldwide. According to Robert, sustainability is fundamentally about maintaining human life on the planet, and thus addressing human needs is an essential element of creating a sustainable society. Complying with the above system conditions should be the aim of organizations, and they must adopt a systems approach to understand their position against each.
The Biosphere Rules	Unruh (2008)	Biosphere rules is a complementary set of principles and the latest redefinition of what is lean and green, which emerged in 2000s. In a nutshell, after 20 years of working in the industry, Unruh observed that the mantra for environmentally responsible materials management—reduce, reuse, recycle—is not as lean as it seems. Manufacturing uses too many basic components and materials to be efficient. Unruh proposes that firms should imitate the lean logic that nature uses to assemble life and structure ecosystems (he calls it bio-logic). In nature, only a few elements are used to create life (C, N, O, H); therefore, when life ends, residues are readily recycled without toxic waste. On the contrary, the industrio-logic of human manufacturing assumes that largely synthetic materials should be assembled or moulded into desired shapes. Unruh suggests three important biosphere rules to be adapted by companies for both environmental and economic gain:

	Use a limited palette of materials
	Cycle up; prepare your product to be easily recycled into a new product
	Exploit the power of platforms or common production systems.
	He presents examples of companies already structuring their production systems following these principles. Although a challenging idea, the approach is too recent to assess its results.

Appendix B

Useful Resources

A number of websites provide detailed information on a variety of lean and green applications; essentially, they offer general tools and resources for carbon/energy efficiency. We suggest the following links.

www.leanandgreenbusiness.com: This is a website dedicated to promote lean and green thinking. There are various resources available to disseminate lean and green knowledge.

Natural Capitalism Solutions, http://www.natcapsolutions.org: Natural Capitalism Solutions (NCS) is a nonprofit organization. NCS is recognized internationally for its work in the field of sustainability. Formed by Hunter Lovins, co-author of the acclaimed book *Natural Capitalism: Creating the Next Industrial Revolution*, Natural Capitalism Solutions is led by Lovins, who has more than forty years experience in business, sustainability, and change management. NCS mission is to educate senior decision makers in business, government, and civil society about the principles of sustainability. NCS shows how to restore and further enhance natural and human capital while increasing prosperity and quality of life. In partnership with leading thinkers and groups, NCS creates innovative, practical tools and implementation strategies for companies, communities, and countries.

Lean Business System, http://www.leanbusinesssystem.com: Lean Business System is a community based on a holistic, sustainable, and contingent approach to lean. It draws on the expertise of a hand-picked global group of thought leaders from academia and industry. It provides a range of resources, many of which are free to access. It provides a glossary of terms that can help if you need to untangle some of the lean terminology. You can work through a personal or a business-oriented learning journey, or seek lean-related advice from Peter Hines and Keivan Zokaei at Doctors' Surgery. You can also watch short videos to find out more about ways you can learn, share, and grow with Lean Business System.

United Kingdom Department for Environment, Food and Rural Affairs, http://www.defra.gov.uk/environment/business/index.html: The United Kingdom Department for Environment, Food and Rural Affairs (DEFRA) offers a specific section on business and the environment in its website to help businesses optimise their environmental performance via practical tips for reducing waste and saving energy. It is explained that the move to a sustainable, low-carbon, and resource-efficient future can potentially offer opportunities for creating economic growth and job creation.

Envirowise, http://www.envirowise.gov.uk/: Envirowise offers free, independent support to businesses to become more resource efficient and save money, a key outcome of successful lean/green business performance. The website contains a number of publications that offer a variety of suggestions to businesses to reduce waste, manage unavoidable waste, and work with suppliers.

Business in the Community, http://www.bitc.org.uk/resources/index.html: Business in the Community has a directory of 150 case studies of responsible business practice classified by type of practice, industrial sector, and geographical area. The case studies are displayed in the Environment section under the subheading Climate Change.

National Industrial Symbiosis Programme, http://www.nisp.network.com: The National Industrial Symbiosis Programme brings together companies from all business sectors with the aim of improving cross-industry resource efficiency through the commercial trading of materials, energy, and water and sharing assets, logistics, and expertise. Its webpage offers free access to case studies, reports, and guidelines.

GreenBiz.com, http://www.greenbiz.com/view-all/tool: GreenBiz.com, the business voice of the green economy, is a profligate source for news, opinion, best practices, and other resources on the greening of mainstream business. The website provides clear, concise, accurate, and balanced information, resources, and learning opportunities to help companies of all sizes and sectors integrate environmental responsibility into their organizations in a manner that supports profitable business practices.

DANTES, http://www.dantes.info/Tools&Methods/Tools_Methods.html: The Digital Agenda for New Tourism Approach (DANTES) website was launched in September 2003 as part of the European Union (EU)-financed DANTES project. It presents a toolbox with many different methods that can be used by, for example, professionals in a company for environmental assessment.

Cleaner Production Germany, http://www.cleaner-production.de/en/: Presented as the gateway for environmental technology, Cleaner Production provides comprehensive information about the performance of German environmental technologies and environmental services.

Natural Resources Canada, http://oee.nrcan.gc.ca/industrial/technical-information.cfm?attr=24: Natural Resources Canada is an organization specialized in earth sciences, forestry, energy, and minerals and metals to promote the responsible and sustainable development of Canada's natural resources. Its website contains practical information for industry, principally to learn how to assess energy efficiency and uncover potential efficiencies.

US Department of Energy, Advanced Manufacturing Office, http://www1.eere.energy.gov/industry/bestpractices/: Established by the US Department of Energy (DOE), this website draws upon the DOE Industrial Technologies Program to propose best practices to implement energy management practices in industrial plants.

Bibliography

Hall, R. *Compression: Meeting the Challenges of Sustainability through Vigorous Learning Enterprises*. New York: Productivity Press, 2009.

Hawken P., A. B. Lovins, and L. H. Lovins. *Natural Capitalism: Creating the Next Industrial Revolution*. London: Earthscan Publications Ltd., 1999.

Hines P., P. Found, G. Griffiths, and R. Harrison. *Staying Lean: Thriving, Not Just Surviving*. Cardiff: Lean Enterprise Research Centre, 2008.

Holcim Foundation. *Clothing Factory in Sri Lanka*, Zurich, Switzerland: Holcim Foundation for Sustainable Construction, 2009.

Leibundgut, H. Clothing factory in Sri Lanka. Zurich: Holcim Foundation for Sustainable Construction, 2009.

Liker, J. *The Toyota Way: 14 Management Principles from the World's Greatest Manufacturer*. New York: McGraw-Hill, 2003.

Zokaei, A. K., J. Seddon, and B. O'Donovan. *Systems Thinking: From Heresy to Practice*, London: Palgrave McMillan, 2011.

Index